P9-BJC-151

Hello
world

www.penguin.co.uk

Hello world

How to Be Human in the Age of the Machine

HANNAH FRY

Doubleday

LONDON · NEW YORK · TORONTO · SYDNEY · AUCKLAND

TRANSWORLD PUBLISHERS
61–63 Uxbridge Road, London W5 5SA
www.penguin.co.uk

Transworld is part of the Penguin Random House group of companies
whose addresses can be found at global.penguinrandomhouse.com

First published in Great Britain in 2018 by Doubleday
an imprint of Transworld Publishers

A CIP catalogue record for this book
is available from the British Library.

ISBNs 9780857525246 (hb)
9780857525253 (tpb)

Typeset in 12/16 pt Minion Pro
by Integra Software Services Pvt. Ltd, Pondicherry

Printed and bound in Great Britain by Clays Ltd, Elcograf S.p.A.

Penguin Random House is committed to a sustainable
future for our business, our readers and our planet. This book
is made from Forest Stewardship Council® certified paper.

9 10 8

For Marie Fry.
Thank you for never taking no for an answer.

Contents

A note on the title

WHEN I WAS 7 YEARS old, my dad brought a gift home for me and my sisters. It was a ZX Spectrum, a little 8-bit computer – the first time we'd ever had one of our own. It was probably already five years out of date by the time it arrived in our house, but even though it was second-hand, I instantly thought there was something marvellous about that dinky machine. The Spectrum was roughly equivalent to a Commodore 64 (although only the really posh kids in the neighbourhood had one of those) but I always thought it was a far more beautiful beast. The sleek black plastic casing could fit in your hands, and there was something rather friendly about the grey rubber keys and rainbow stripe running diagonally across one corner.

For me, the arrival of that ZX Spectrum marked the beginning of a memorable summer spent up in the loft with my elder sister, programming hangman puzzles for each other, or drawing simple shapes through code. All that 'advanced' stuff came later, though. First we had to master the basics.

Looking back, I don't exactly remember the moment I wrote my first ever computer program, but I'm pretty sure I know what it was. It would have been the same simple program that I've gone on to teach all of my students at University College London; the same as you'll find on the first page of practically any introductory computer science textbook. Because there is a tradition among all those who have ever learned to code – a rite of passage, almost. Your first task as a rookie is to program the computer to flash up a famous phrase on to the screen:

'HELLO WORLD'

It's a tradition that dates back to the 1970s, when Brian Kernighan included it as a tutorial in his phenomenally popular programming textbook.[1] The book – and hence the phrase – marked an important point in the history of computers. The microprocessor had just arrived on the scene, heralding the transition of computers from what they had been in the past – enormous great specialist machines, fed on punch cards and ticker-tape – to something more like the personal computers we're used to, with a screen, a keyboard and a blinking cursor. 'Hello world' came along at the first moment when chit-chat with your computer was a possibility.

Years later, Brian Kernighan told a *Forbes* interviewer about his inspiration for the phrase. He'd seen a cartoon showing an egg and a newly hatched chick chirping the words 'Hello world!' as it was born, and it had stuck in his mind.

It's not entirely clear who the chick is supposed to be in that scenario: the fresh-faced human triumphantly announcing their brave arrival to the world of programming? Or the computer itself, awakening from the mundane slumber of spreadsheets and text documents, ready to connect its mind to the real world and do its new master's bidding? Maybe both. But it's certainly a phrase that unites all programmers, and connects them to every machine that's ever been programmed.

There's something else I like about the phrase – something that has never been more relevant or more important than it is now. As computer algorithms increasingly control and decide our future, 'Hello world' is a reminder of a moment of dialogue between human and machine. Of an instant where the boundary between controller and controlled is virtually imperceptible. It marks the start of a partnership – a shared journey of possibilities, where one cannot exist without the other.

In the age of the machine, that's a sentiment worth bearing in mind.

Introduction

ANYONE WHO HAS EVER VISITED Jones Beach on Long Island, New York, will have driven under a series of bridges on their way to the ocean. These bridges, primarily built to filter people on and off the highway, have an unusual feature. As they gently arc over the traffic, they hang extraordinarily low, sometimes leaving as little as 9 feet of clearance from the tarmac.

There's a reason for this strange design. In the 1920s, Robert Moses, a powerful New York urban planner, was keen to keep his newly finished, award-winning state park at Jones Beach the preserve of white and wealthy Americans. Knowing that his preferred clientele would travel to the beach in their private cars, while people from poor black neighbourhoods would get there by bus, he deliberately tried to limit access by building hundreds of low-lying bridges along the highway. Too low for the 12-foot buses to pass under.[1]

Racist bridges aren't the only inanimate objects that have had a quiet, clandestine control over people. History is littered with

1

examples of objects and inventions with a power beyond their professed purpose.[2] Sometimes it's deliberately and maliciously factored into their design, but at other times it's a result of thought-less omissions: just think of the lack of wheelchair access in some urban areas. Sometimes it's an unintended consequence, like the mechanized weaving machines of the nineteenth century. They were designed to make it easier to create complicated textiles, but in the end, the impact they had on wages, unemployment and working conditions made them arguably more tyrannical than any Victorian capitalist.

Modern inventions are no different. Just ask the residents of Scunthorpe in the north of England, who were blocked from open-ing AOL accounts after the internet giant created a new profanity filter that objected to the name of their town.[3] Or Chukwuemeka Afigbo, the Nigerian man who discovered an automatic hand-soap dispenser that released soap perfectly whenever his white friend placed their hand under the machine, but refused to acknowledge his darker skin.[4] Or Mark Zuckerberg, who, when writing the code for Facebook in his dorm room in Harvard in 2004, would never have imagined his creation would go on to be accused of helping manipulate votes in elections around the globe.[5]

Behind each of these inventions is an algorithm. The invisible pieces of code that form the gears and cogs of the modern machine age, algorithms have given the world everything from social media feeds to search engines, satellite navigation to music recommenda-tion systems, and are as much a part of our modern infrastructure as bridges, buildings and factories ever were. They're inside our hos-pitals, our courtrooms and our cars. They're used by police forces, supermarkets and film studios. They have learned our likes and dislikes; they tell us what to watch, what to read and who to date. And all the while, they have the hidden power to slowly and subtly change the rules about what it means to be human.

In this book, we'll discover the vast array of algorithms on which we increasingly, but perhaps unknowingly, rely. We'll pay close attention to their claims, examine their undeclared power and confront the unanswered questions they raise. We'll encounter algorithms used by police to decide who should be arrested, which make us choose between protecting the victims of crime and the innocence of the accused. We'll meet algorithms used by judges to decide on the sentences of convicted criminals, which ask us to decide what our justice system should look like. We'll find algorithms used by doctors to over-rule their own diagnoses; algorithms within driverless cars that insist we define our morality; algorithms that are weighing in on our expressions of emotion; and algorithms with the power to undermine our democracies.

I'm not arguing that algorithms are inherently bad. As you'll see in these pages, there are many reasons to be positive and optimistic about what lies ahead. No object or algorithm is ever either good or evil in itself. It's how they're used that matters. GPS was invented to launch nuclear missiles and now helps deliver pizzas. Pop music, played on repeat, has been deployed as a torture device. And however beautifully made a garland of flowers might be, if I really wanted to I could strangle you with it. Forming an opinion on an algorithm means understanding the relationship between human and machine. Each one is inextricably connected to the people who build and use it.

This means that, at its heart, this is a book about humans. It's about who we are, where we're going, what's important to us and how that is changing through technology. It's about our relationship with the algorithms that are already here, the ones working alongside us, amplifying our abilities, correcting our mistakes, solving our problems and creating new ones along the way.

It's about asking if an algorithm is having a net benefit on society. About when you should trust a machine over your own judgement, and when you should resist the temptation to leave machines in control. It's about breaking open the algorithms and finding their limits; and about looking hard at ourselves and finding our own. About separating the harm from the good and deciding what kind of world we want to live in.

Because the future doesn't just happen. We create it.

Power

GARRY KASPAROV KNEW EXACTLY HOW to intimidate his rivals. At 34, he was the greatest chess player the world had ever seen, with a reputation fearsome enough to put any opponent on edge. Even so, there was one unnerving trick in particular that his competitors had come to dread. As they sat, sweating through what was probably the most difficult game of their life, the Russian would casually pick up his watch from where it had been lying beside the chessboard, and return it to his wrist. This was a signal that everybody recognized – it meant that Kasparov was bored with toying with his opponent. The watch was an instruction that it was time for his rival to resign the game. They could refuse, but either way, Kasparov's victory was soon inevitable.[1]

But when IBM's Deep Blue faced Kasparov in the famous match of May 1997, the machine was immune to such tactics. The outcome of the match is well known, but the story behind how Deep Blue secured its win is less widely appreciated. That symbolic victory, of machine over man, which in many ways marked the start of the algorithmic age, was down to far more than sheer raw computing power. In order to beat Kasparov, Deep Blue had to understand him not simply as a highly efficient processor of brilliant chess moves, but as a human being.

For a start, the IBM engineers made the brilliant decision to design Deep Blue to appear more uncertain than it was. During their infamous six-game match, the machine would occasionally hold off from declaring its move once a calculation had finished, sometimes for several minutes. From Kasparov's end of the table, the delays made it look as if the machine was struggling, churning through more and more calculations. It seemed to confirm what Kasparov thought he knew; that he'd successfully dragged the game into a position where the number of possibilities was so mind-bogglingly large that Deep Blue couldn't make a sensible decision.[2] In reality, however, it was sitting idly by, knowing exactly what to play, just letting the clock tick down. It was a mean trick, but it worked. Even in the first game of the match, Kasparov started to become distracted by second-guessing how capable the machine might be.[3]

Although Kasparov won the first game, it was in game two that Deep Blue really got into his head. Kasparov tried to lure the computer into a trap, tempting it to come in and capture some pieces, while at the same time setting himself up – several moves ahead – to release his queen and launch an attack.[4] Every watching chess expert expected the computer to take the bait, as did Kasparov himself. But somehow, Deep Blue smelt a rat. To Kasparov's amazement, the computer had realized what the grandmaster was planning and moved to block his queen, killing any chance of a human victory.[5]

Kasparov was visibly horrified. His misjudgement about what the computer could do had thrown him. In an interview a few days after the match he described Deep Blue as having 'suddenly played like a god for one moment'.[6] Many years later, reflecting on how he had felt at the time, he would write that he had 'made the mistake of assuming that moves that were surprising for a computer to make were also objectively strong moves'.[7] Either way, the genius of the algorithm had triumphed. Its understanding of the human mind, and human fallibility, was attacking and defeating the all-too-human genius.

Disheartened, Kasparov resigned the second game rather than fighting for the draw. From there his confidence began to unravel.

Games three, four and five ended in draws. By game six, Kasparov was broken. The match ended Deep Blue 3½ to Kasparov's 2½.

It was a strange defeat. Kasparov was more than capable of working his way out of those positions on the board, but he had underestimated the ability of the algorithm and then allowed himself to be intimidated by it. 'I had been so impressed by Deep Blue's play,' he wrote in 2017, reflecting on the match. 'I became so concerned with what it might be capable of that I was oblivious to how my problems were more due to how badly I was playing than how well it was playing.'[8]

As we'll see time and time again in this book, expectations are important. The story of Deep Blue defeating the great grandmaster demonstrates that the power of an algorithm isn't limited to what is contained within its lines of code. Understanding our own flaws and weaknesses – as well as those of the machine – is the key to remaining in control.

But if someone like Kasparov failed to grasp this, what hope is there for the rest of us? Within these pages, we'll see how algorithms have crept into virtually every aspect of modern life – from health and crime to transport and politics. Along the way, we have somehow managed to be simultaneously dismissive of them, intimidated by them and in awe of their capabilities. The end result is that we have no idea quite how much power we're ceding, or if we've let things go too far.

Back to basics

Before we get to all that, perhaps it's worth pausing briefly to question what 'algorithm' actually means. It's a term that, although used frequently, routinely fails to convey much actual information. This is partly because the word itself is quite vague. Officially, it is defined as follows:[9]

> *algorithm* (noun): A step-by-step procedure for solving a problem or accomplishing some end especially by a computer.

7

That's it. An algorithm is simply a series of logical instructions that show, from start to finish, how to accomplish a task. By this broad definition, a cake recipe counts as an algorithm. So does a list of directions you might give to a lost stranger. IKEA manuals, YouTube troubleshooting videos, even self-help books – in theory, any self-contained list of instructions for achieving a specific, defined objective could be described as an algorithm.

But that's not quite how the term is used. Usually, algorithms refer to something a little more specific. They still boil down to a list of step-by-step instructions, but these algorithms are almost always mathematical objects. They take a sequence of mathematical operations – using equations, arithmetic, algebra, calculus, logic and probability – and translate them into computer code. They are fed with data from the real world, given an objective and set to work crunching through the calculations to achieve their aim. They are what makes computer science an actual science, and in the process have fuelled many of the most miraculous modern achievements made by machines.

There's an almost uncountable number of different algorithms. Each has its own goals, its own idiosyncrasies, its clever quirks and drawbacks, and there is no consensus on how best to group them. But broadly speaking, it can be useful to think of the real-world tasks they perform in four main categories:[10]

Prioritization: making an ordered list

Google Search predicts the page you're looking for by ranking one result over another. Netflix suggests which films you might like to watch next. Your TomTom selects your fastest route. All use a mathematical process to order the vast array of possible choices. Deep Blue was also essentially a prioritization algorithm, reviewing all the possible moves on the chessboard and calculating which would give the best chance of victory.

Classification: picking a category

As soon as I hit my late twenties, I was bombarded by adverts for diamond rings on Facebook. And once I eventually got married, adverts for pregnancy tests followed me around the internet. For these mild irritations, I had classification algorithms to thank. These algorithms, loved by advertisers, run behind the scenes and classify you as someone interested in those things on the basis of your characteristics. (They might be right, too, but it's still annoying when adverts for fertility kits pop up on your laptop in the middle of a meeting.)

There are algorithms that can automatically classify and remove inappropriate content on YouTube, algorithms that will label your holiday photos for you, and algorithms that can scan your handwriting and classify each mark on the page as a letter of the alphabet.

Association: finding links

Association is all about finding and marking relationships between things. Dating algorithms such as OKCupid have association at their core, looking for connections between members and suggesting matches based on the findings. Amazon's recommendation engine uses a similar idea, connecting your interests to those of past customers. It's what led to the intriguing shopping suggestion that confronted Reddit user Kerbobotat after buying a baseball bat on Amazon: 'Perhaps you'll be interested in this balaclava?'[11]

Filtering: isolating what's important

Algorithms often need to remove some information to focus on what's important, to separate the signal from the noise. Sometimes they do this literally: speech recognition algorithms, like those running inside Siri, Alexa and Cortana, first need to filter out your voice from the background noise before they can get to work on deciphering what you're saying. Sometimes they do it figuratively: Facebook and Twitter filter stories that relate to your known interests to design your own personalized feed.

The vast majority of algorithms will be built to perform a combination of the above. Take UberPool, for instance, which matches prospective passengers with others heading in the same direction. Given your start point and end point, it has to filter through the possible routes that could get you home, look for connections with other users headed in the same direction, and pick one group to assign you to – all while prioritizing routes with the fewest turns for the driver, to make the ride as efficient as possible.[12]

So, that's what algorithms can do. Now, how do they manage to do it? Well, again, while the possibilities are practically endless, there is a way to distil things. You can think of the approaches taken by algorithms as broadly fitting into two key paradigms, both of which we'll meet in this book.

Rule-based algorithms

The first type are rule-based. Their instructions are constructed by a human and are direct and unambiguous. You can imagine these algorithms as following the logic of a cake recipe. Step one: do this. Step two: if this, then that. That's not to imply that these algorithms are simple – there's plenty of room to build powerful programs within this paradigm.

Machine-learning algorithms

The second type are inspired by how living creatures learn. To give you an analogy, think about how you might teach a dog to give you a high five. You don't need to produce a precise list of instructions and communicate them to the dog. As a trainer, all you need is a clear objective in your mind of what you want the dog to do and some way of rewarding her when she does the right thing. It's simply about reinforcing good behaviour, ignoring bad, and giving her enough practice to work out what to do for herself. The algorithmic equivalent is known as a *machine-learning algorithm*, which comes under the broader umbrella of *artificial intelligence* or *AI*. You give the machine data, a goal and feedback when it's on the right track – and leave it to work out the best way of achieving the end.

Both types have their pros and cons. Because rule-based algorithms have instructions written by humans, they're easy to comprehend. In theory, anyone can open them up and follow the logic of what's happening inside.[13] But their blessing is also their curse. Rule-based algorithms will only work for the problems for which humans know how to write instructions.

Machine-learning algorithms, by contrast, have recently proved to be remarkably good at tackling problems where writing a list of instructions won't work. They can recognize objects in pictures, understand words as we speak them and translate from one language to another – something rule-based algorithms have always struggled with. The downside is that if you let a machine figure out the solution for itself, the route it takes to get there often won't make a lot of sense to a human observer. The insides can be a mystery, even to the smartest of living programmers.

Take, for instance, the job of image recognition. A group of Japanese researchers recently demonstrated how strange an algorithm's way of looking at the world can seem to a human. You might have come across the optical illusion where you can't quite tell if you're looking at a picture of a vase or of two faces (if not, there's an example in the notes at the back of the book).[14] Here's the computer equivalent. The team showed that changing a single pixel on the front wheel of the image overleaf was enough to cause a machine-learning algorithm to change its mind from thinking this is a photo of a car to thinking it is a photo of a dog.[15]

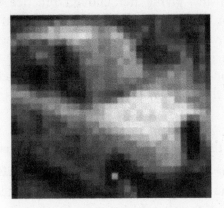

For some, the idea of an algorithm working without explicit instructions is a recipe for disaster. How can we control something we don't understand? What if the capabilities of sentient, super-intelligent machines transcend those of their makers? How will we ensure that an AI we don't understand and can't control isn't working against us?

These are all interesting hypothetical questions, and there is no shortage of books dedicated to the impending threat of an AI apocalypse. Apologies if that was what you were hoping for, but this book isn't one of them. Although AI has come on in leaps and bounds of late, it is still only 'intelligent' in the narrowest sense of the word. It would probably be more useful to think of what we've been through as a revolution in computational statistics than a revolution in intelligence. I know that makes it sound a lot less sexy (unless you're *really* into statistics), but it's a far more accurate description of how things currently stand.

For the time being, worrying about evil AI is a bit like worrying about overcrowding on Mars.* Maybe one day we'll get to the point

* This is paraphrased from a comment made by the computer scientist and machine-learning pioneer Andrew Ng in a talk he gave in 2015. See Tech Events, 'GPU Technology Conference 2015 day 3: What's Next in Deep Learning', *YouTube*, 20 Nov. 2015, https://www.youtube.com/watch?v=qP9TOX8T-kI.

where computer intelligence surpasses human intelligence, but we're nowhere near it yet. Frankly, we're still quite a long way away from creating hedgehog-level intelligence. So far, no one's even managed to get past worm.*

Besides, all the hype over AI is a distraction from much more pressing concerns and – I think – much more interesting stories. Forget about omnipotent artificially intelligent machines for a moment and turn your thoughts from the far distant future to the here and now – because there are already algorithms with free rein to act as autonomous decision-makers. To decide prison terms, treatments for cancer patients and what to do in a car crash. They're already making life-changing choices on our behalf at every turn.

The question is, if we're handing over all that power – are they deserving of our trust?

Blind faith

Sunday, 22 March 2009 wasn't a good day for Robert Jones. He had just visited some friends and was driving back through the pretty town of Todmorden in West Yorkshire when he noticed the fuel light on his BMW. He had just 7 miles to find a petrol station before he ran out, which was cutting things rather fine. Thankfully his GPS seemed to have found him a short cut – sending him on a narrow winding path up the side of the valley.

Robert followed the machine's instructions, but as he drove, the road got steeper and narrower. After a couple of miles, it turned into a dirt track that barely seemed designed to accommodate

---

* Simulating the brain of a worm is precisely the goal of the international science project OpenWorm. They're hoping to artificially reproduce the network of 302 neurons found within the brain of the *C. elegans* worm. To put that into perspective, we humans have around 100,000,000,000 neurons. See OpenWorm website: http://openworm.org/.

horses, let alone cars. But Robert wasn't fazed. He drove five thousand miles a week for a living and knew how to handle himself behind the wheel. Plus, he thought, he had 'no reason not to trust the TomTom sat-nav'.[16]

Just a short while later, anyone who happened to be looking up from the valley below would have seen the nose of Robert's BMW appearing over the brink of the cliff above, saved from the hundred-foot drop only by the flimsy wooden fence at the edge he'd just crashed into.

It would eventually take a tractor and three quad bikes to recover Robert's car from where he abandoned it. Later that year, when he appeared in court on charges of reckless driving, he admitted that he didn't think to over-rule the machine's instructions. 'It kept insisting the path was a road,' he told a newspaper after the incident. 'So I just trusted it. You don't expect to be taken nearly over a cliff.'[17]

No, Robert. I guess you don't.

There's a moral somewhere in this story. Although he probably felt a little foolish at the time, in ignoring the information in front of his eyes (like seeing a sheer drop out of the car window) and attributing greater intelligence to an algorithm than it deserved, Jones was in good company. After all, Kasparov had fallen into the same trap some twelve years earlier. And, in much quieter but no less profound ways, it's a mistake almost all of us are guilty of making, perhaps without even realizing.

Back in 2015 scientists set out to examine how search engines like Google have the power to alter our view of the world.[18] They wanted to find out if we have healthy limits in the faith we place in their results, or if we would happily follow them over the edge of a metaphorical cliff.

The experiment focused around an upcoming election in India. The researchers, led by psychologist Robert Epstein, recruited

2,150 undecided voters from around the country and gave them access to a specially made search engine, called 'Kadoodle', to help them learn more about the candidates before deciding who they would vote for.

Kadoodle was rigged. Unbeknown to the participants, they had been split into groups, each of which was shown a slightly different version of the search engine results, biased towards one candidate or another. When members of one group visited the website, all the links at the top of the page would favour one candidate in particular, meaning they'd have to scroll right down through link after link before finally finding a single page that was favourable to anyone else. Different groups were nudged towards different candidates.

It will come as no surprise that the participants spent most of their time reading the websites flagged up at the top of the first page – as that old internet joke says, the best place to hide a dead body is on the second page of Google search results. Hardly anyone in the experiment paid much attention to the links that appeared well down the list. But still, the degree to which the ordering influenced the volunteers' opinions shocked even Epstein. After only a few minutes of looking at the search engine's biased results, when asked who they would vote for, participants were a staggering 12 per cent more likely to pick the candidate Kadoodle had favoured.

In an interview with *Science* in 2015,[19] Epstein explained what was going on: 'We expect the search engine to be making wise choices. What they're saying is, "Well yes, I see the bias and that's telling me ... the search engine is doing its job."' Perhaps more ominous, given how much of our information we now get from algorithms like search engines, is how much agency people believed they had in their own opinions: 'When people are unaware they are being manipulated, they tend to believe they have adopted their new thinking voluntarily,' Epstein wrote in the original paper.[20]

Kadoodle, of course, is not the only algorithm to have been accused of subtly manipulating people's political opinions. We'll come on to that more in the 'Data' chapter, but for now it's worth noting how the experiment suggests we feel about algorithms that are right most of the time. We end up believing that they always have superior judgement.[21] After a while, we're no longer even aware of our own bias towards them.

All around us, algorithms provide a kind of convenient source of authority. An easy way to delegate responsibility; a short cut that we take without thinking. Who is really going to click through to the second page of Google every time and think critically about every result? Or go to every airline to check if Skyscanner is listing the cheapest deals? Or get out a ruler and a road map to confirm that their GPS is offering the shortest route? Not me, that's for sure.

But there's a distinction that needs making here. Because trusting a usually reliable algorithm is one thing. Trusting one without any firm understanding of its quality is quite another.

Artificial intelligence meets natural stupidity

In 2012, a number of disabled people in Idaho were informed that their Medicaid assistance was being cut.[22] Although they all qualified for benefits, the state was slashing their financial support – without warning – by as much as 30 per cent,[23] leaving them struggling to pay for their care. This wasn't a political decision; it was the result of a new 'budget tool' that had been adopted by the Idaho Department of Health and Welfare – a piece of software that automatically calculated the level of support that each person should receive.[24]

The problem was, the budget tool's decisions didn't seem to make much sense. As far as anyone could tell from the outside, the numbers it came up with were essentially arbitrary. Some people were given more money than in previous years, while others found their budgets reduced by tens of thousands of dollars, putting them at risk of having to leave their homes to be cared for in an institution.[25]

Unable to understand why their benefits had been reduced, or to effectively challenge the reduction, the residents turned to the American Civil Liberties Union (ACLU) for help. Their case was taken on by Richard Eppink, legal director of the Idaho division,[26] who had this to say in a blog post in 2017: 'I thought the case would be a simple matter of saying to the state: Okay, tell us why these dollar figures dropped by so much?'[27] In fact, it would take four years, four thousand plaintiffs and a class action lawsuit to get to the bottom of what had happened.[28]

Eppink and his team began by asking for details on how the algorithm worked, but the Medicaid team refused to explain their calculations. They argued that the software that assessed the cases was a 'trade secret' and couldn't be shared.[29] Fortunately, the judge presiding over the case disagreed. The budget tool that wielded so much power over the residents was then handed over, and revealed to be – not some sophisticated AI, not some beautifully crafted mathematical model, but an Excel spreadsheet.[30]

Within the spreadsheet, the calculations were supposedly based on historical cases, but the data was so badly riddled with bugs and errors that it was, for the most part, entirely useless.[31] Worse, once the ACLU team managed to unpick the equations, they discovered 'fundamental statistical flaws in the way that the formula itself was structured'. The budget tool had effectively been producing random results for a huge number of people. The algorithm – if you can call it that – was of such poor quality that the court would eventually rule it unconstitutional.[32]

There are two parallel threads of human error here. First, someone wrote this garbage spreadsheet; second, others naïvely trusted it. The 'algorithm' was in fact just shoddy human work wrapped up in code. So why were the people who worked for the state so eager to defend something so terrible?

Here are Eppink's thoughts on the matter:

> It's just this bias we all have for computerized results – we don't question them. When a computer generates something – when you have a statistician, who looks at some data, and comes up with a formula – we just trust that formula, without asking 'hey wait a second, how is this actually working?'[33]

Now, I know that picking mathematical formulae apart to see how they work isn't everyone's favourite pastime (even if it is mine). But Eppink none the less raises an incredibly important point about our human willingness to take algorithms at face value without wondering what's going on behind the scenes.

In my years working as a mathematician with data and algorithms, I've come to believe that the only way to objectively judge whether an algorithm is trustworthy is by getting to the bottom of how it works. In my experience, algorithms are a lot like magical illusions. At first they appear to be nothing short of actual wizardry, but as soon as you know how the trick is done, the mystery evaporates. Often there's something laughably simple (or worryingly reckless) hiding behind the façade. So, in the chapters that follow, and the algorithms we'll explore, I'll try to give you a flavour of what's going on behind the scenes where I can. Enough to see how the tricks are done – even if not quite enough to perform them yourself.

But even for the most diehard maths fans, there are still going to be occasions where algorithms demand you take a blind leap of faith. Perhaps because, as with Skyscanner or Google's search results, double-checking their working isn't feasible. Or maybe, like the Idaho budget tool and others we'll meet, the algorithm is considered a 'trade secret'. Or perhaps, as in some machine-learning techniques, following the logical process inside the algorithm just isn't possible.

There will be times when we have to hand over control to the unknown, even while knowing that the algorithm is capable of making mistakes. Times when we are forced to weigh up our own judgement against that of the machine. When, if we decide to trust our instincts instead of its calculations, we're going to need rather a lot of courage in our convictions.

When to over-rule

Stanislav Petrov was a Russian military officer in charge of monitoring the nuclear early warning system protecting Soviet airspace. His job was to alert his superiors immediately if the computer indicated any sign of an American attack.[34]

Petrov was on duty on 26 September 1983 when, shortly after midnight, the sirens began to howl. This was the alert that everyone dreaded. Soviet satellites had detected an enemy missile headed for Russian territory. This was the depths of the Cold War, so a strike was certainly plausible, but something gave Petrov pause. He wasn't sure he trusted the algorithm. It had only detected five missiles, which seemed like an illogically small opening salvo for an American attack.[35]

Petrov froze in his chair. It was down to him: report the alert, and send the world into almost certain nuclear war; or wait, ignoring protocol, knowing that with every second that passed his country's leaders had less time to launch a counter-strike.

Fortunately for all of us, Petrov chose the latter. He had no way of knowing for sure that the alarm had sounded in error, but after 23 minutes – which must have felt like an eternity at the time – when it was clear that no nuclear missiles had landed on Russian soil, he finally knew that he had been correct. The algorithm had made a mistake.

If the system had been acting entirely autonomously, without a human like Petrov to act as the final arbiter, history would undoubtedly have played out rather differently. Russia would almost certainly have launched what it believed to be retaliatory action and triggered a full-blown nuclear war in the process. If there's anything we can learn from this story, it's that the human element does seem to be a critical part of the process: that having a person with the power of veto in a position to review the suggestions of an algorithm before a decision is made is the only sensible way to avoid mistakes.

After all, only humans will feel the weight of responsibility for their decisions. An algorithm tasked with communicating up to the Kremlin wouldn't have thought for a second about the potential ramifications of such a decision. But Petrov, on the other hand? 'I knew perfectly well that nobody would be able to correct my mistake if I had made one.'[36]

The only problem with this conclusion is that humans aren't always that reliable either. Sometimes, like Petrov, they'll be right to over-rule an algorithm. But often our instincts are best ignored.

To give you another example from the world of safety, where stories of humans incorrectly over-ruling an algorithm are mercifully rare, that is none the less precisely what happened during an infamous crash on the Smiler rollercoaster at Alton Towers, the UK's biggest theme park.[37]

Back in June 2015, two engineers were called to attend a fault on a rollercoaster. After fixing the issue, they sent an empty carriage around to test everything was working – but failed to notice it never made it back. For whatever reason, the spare carriage rolled backwards down an incline and came to a halt in the middle of the track.

20

Meanwhile, unbeknown to the engineers, the ride staff added an extra carriage to deal with the lengthening queues. Once they got the all-clear from the control room, they started loading up the carriages with cheerful passengers, strapping them in and sending the first car off around the track, completely unaware of the empty, stranded carriage sent out by the engineers sitting directly in its path.

Luckily, the rollercoaster designers had planned for a situation like this, and their safety algorithms worked exactly as planned. To avoid a certain collision, the packed train was brought to a halt at the top of the first climb, setting off an alarm in the control room. But the engineers – confident that they'd just fixed the ride – concluded the automatic warning system was at fault.

Over-ruling the algorithm wasn't easy: they both had to agree and simultaneously press a button to restart the rollercoaster. Doing so sent the train full of people over the drop to crash straight into the stranded extra carriage. The result was horrendous. Several people suffered devastating injuries and two teenage girls lost their legs.

Both of these life-or-death scenarios, Alton Towers and Petrov's alarm, serve as dramatic illustrations of a much deeper dilemma. In the balance of power between human and algorithm, who – or what – should have the final say?

Power struggle

This is a debate with a long history. In 1954, Paul Meehl, a professor of clinical psychology at the University of Minnesota, annoyed an entire generation of humans when he published *Clinical versus Statistical Prediction*, coming down firmly on one side of the argument.[38]

In his book, Meehl systematically compared the performance of humans and algorithms on a whole variety of subjects – predicting everything from students' grades to patients' mental health outcomes – and concluded that mathematical algorithms, no matter how simple, will almost always make better predictions than people.

Countless other studies in the half-century since have confirmed Meehl's findings. If your task involves any kind of calculation, put your money on the algorithm every time: in making medical diagnoses or sales forecasts, predicting suicide attempts or career satisfaction, and assessing everything from fitness for military service to projected academic performance.[39] The machine won't be perfect, but giving a human a veto over the algorithm would just add more error.*

Perhaps this shouldn't come as a surprise. We're not built to compute. We don't go to the supermarket to find a row of cashiers eyeballing our shopping to gauge how much it should cost. We get an (incredibly simple) algorithm to calculate it for us instead. And most of the time, we'd be better off leaving the machine to it. It's like the saying among airline pilots that the best flying team has three components: a pilot, a computer and a dog. The computer is there to fly the plane, the pilot is there to feed the dog. And the dog is there to bite the human if it tries to touch the computer.

But there's a paradox in our relationship with machines. While we have a tendency to over-trust anything we don't understand, as soon as we *know* an algorithm can make mistakes, we also have a rather annoying habit of over-reacting and dismissing it completely, reverting instead to our own flawed judgement. It's known

---

* Intriguingly, a rare exception to the superiority of algorithmic performance comes from a selection of studies conducted in the late 1950s and 1960s into the 'diagnosis' (their words, not mine) of homosexuality. In those examples, the human judgement made far better predictions, outperforming anything the algorithm could manage – suggesting there are some things so intrinsically human that data and mathematical formulae will always struggle to describe them.

to researchers as *algorithm aversion*. People are less tolerant of an algorithm's mistakes than of their own – even if their own mistakes are bigger.

It's a phenomenon that has been demonstrated time and time again in experiments,[40] and to some extent, you might recognize it in yourself. Whenever Citymapper says my journey will take longer than I expect it to, I always think I know better (even if most of the time it means I end up arriving late). We've all called Siri an idiot at least once, somehow in the process forgetting the staggering technological accomplishment it has taken to build a talking assistant you can hold in your hand. And in the early days of using the mobile GPS app Waze I'd find myself sitting in a traffic jam, having been convinced that taking the back roads would be faster than the route shown. (It almost always wasn't.) Now I've come to trust it and – like Robert Jones and his BMW – I'll blindly follow it wherever it leads me (although I still think I'd draw the line at going over a cliff).

This tendency of ours to view things in black and white – seeing algorithms as either omnipotent masters or a useless pile of junk – presents quite a problem in our high-tech age. If we're going to get the most out of technology, we're going to need to work out a way to be a bit more objective. We need to learn from Kasparov's mistake and acknowledge our own flaws, question our gut reactions and be a bit more aware of our feelings towards the algorithms around us. On the flip side, we should take algorithms off their pedestal, examine them a bit more carefully and ask if they're really capable of doing what they claim. That's the only way to decide if they deserve the power they've been given.

Unfortunately, all this is often much easier said than done. Oftentimes, we'll have little say over the power and reach of the algorithms that surround us, even when it comes to those that affect us directly.

This is particularly true for the algorithms that trade in the most fundamental modern commodity: data. The algorithms that silently follow us around the internet, the ones that are harvesting our personal information, invading our privacy and inferring our character with free rein to subtly influence our behaviour. In that perfect storm of misplaced trust and power and influence, the consequences have the potential to fundamentally alter our society.

Data

Back in 2004, soon after college student Mark Zuckerberg created Facebook, he had an instant messenger exchange with a friend:

> Zuck: Yeah so if you ever need info about anyone at Harvard
> Zuck: Just ask.
> Zuck: I have over 4,000 emails, pictures, addresses . . .
> [Redacted friend's name]: What? How'd you manage that one?
> Zuck: People just submitted it.
> Zuck: I don't know why.
> Zuck: They 'trust me'
> Zuck: Dumb fucks[1]

In the wake of the 2018 Facebook scandal, these words were repeatedly reprinted by journalists wanting to hint at a Machiavellian attitude to privacy within the company. Personally, I think we can be a little more generous when interpreting the boastful comments of a 19-year-old. But I also think that Zuckerberg is wrong. People weren't just giving him their details. They were submitting them as part of an exchange. In return, they were given access to an algorithm that would let them freely connect with friends and family, a space to share their lives with others. Their own private network in the vastness of the World Wide Web. I don't know about you, but at the time I certainly thought that was a fair swap.

There's just one issue with that logic: we're not always aware of the longer-term implications of that trade. It's rarely obvious what our data can do, or, when fed into a clever algorithm, just how valuable it can be. Nor, in turn, how cheaply we were bought.

Every little helps

Supermarkets were among the first to recognize the value of an individual's data. In a sector where companies are continually fighting for the customer's attention – for tiny margins of preference that will nudge people's buying behaviour into loyalty to their brand – every slim improvement can add up to an enormous advantage. This was the motivation behind a ground-breaking trial run in 1993 by the British supermarket Tesco.

Under the guidance of husband-and-wife team Edwina Dunn and Clive Humby, and beginning in certain selected stores, Tesco released its brand-new Clubcard – a plastic card, the size and shape of a credit card, that customers could present at a checkout when paying for their shopping. The exchange was simple. For each transaction using a Clubcard, the customer would collect points that they could use against future purchases in store, while Tesco would take a record of the sale and associate it with the customer's name.[2]

The data gathered in that first Clubcard trial was extremely limited. Along with the customer's name and address, the scheme only recorded what they spent and when, not which items were in their basket. None the less, from this modest harvest of data Dunn and Humby reaped some phenomenally valuable insights.

They discovered that a small handful of loyal customers accounted for a massive amount of their sales. They saw, postcode by postcode, how far people were willing to travel to their stores. They uncovered the neighbourhoods where the competition was winning and neighbourhoods where Tesco had the upper hand. The data revealed which customers came back day after day, and which saved their shopping for weekends. Armed with that knowledge, they could get to work nudging their customers' buying behaviour, by sending out a series of coupons to the Clubcard users in the post. High spenders were given vouchers ranging from £3 to £30. Low spenders were sent a smaller incentive of £1 to £10. And the results were staggering. Nearly 70 per cent of the coupons were redeemed, and while in the stores, customers filled up their baskets: people who had Clubcards spent 4 per cent more overall than those who didn't.

On 22 November 1994, Clive Humby presented the findings from the trial to the Tesco board. He showed them the data, the response rates, the evidence of customer satisfaction, the sales boosts. The board listened in silence. At the end of the presentation, the chair was the first person to speak. 'What scares me about this,' he said, 'is that you know more about my customers in three months than I know in 30 years.'[3]

Clubcard was rolled out to all customers of Tesco and is widely credited with putting the company ahead of its main rival Sainsbury's, to become the biggest supermarket in the UK. As time wore on, the data collected became more detailed, making customers' buying habits easier to target.

Early in the days of online shopping, the team introduced a feature known as 'My Favourites', in which any items that were bought while using the loyalty card would appear prominently when the customer logged on to the Tesco website. Like the Clubcard itself, the feature was a roaring success. People could quickly find the products they wanted without having to navigate through the various pages. Sales went up, customers were happy.

But not all of them. Shortly after the launch of the feature, one woman contacted Tesco to complain that her data was wrong. She'd been shopping online and seen condoms among her list of 'My Favourites'. They couldn't be her husband's, she explained, because he didn't use them. At her request, the Tesco analysts looked into the data and discovered that her list was accurate. However, rather than be the cause of a marital rift, they took the diplomatic decision to apologize for 'corrupted data' and remove the offending items from her favourites.

According to Clive Humby's book on Tesco, this has now become an informal policy within the company. Whenever something comes up that is just a bit too revealing, they apologize and delete the data. It's a stance that's echoed by Eric Schmidt, who, while serving as the executive chairman of Google, said he tries to think of things in terms of an imaginary creepy line. 'The Google policy is to get right up to the creepy line but not cross it.'[4]

But collect enough data and it's hard to know what you'll uncover. Groceries aren't just what you consume. They're personal. Look carefully enough at someone's shopping habits and they'll often reveal all kinds of detail about who they are as a person. Sometimes – as in the case of the condoms – it'll be things you'd rather not know. But more often than not, lurking deep within the data, those slivers of hidden insight can be used to a company's advantage.

Target market

Back in 2002, the American discount superstore Target started looking for unusual patterns in its data.[5] Target sells everything from milk and bananas to cuddly toys and garden furniture, and – like pretty much every other retailer since the turn of the millennium – has ways of using credit card numbers and survey responses to tie customers to everything they've ever bought in the store, enabling them to analyse what people are buying.

In a story that – as US readers won't need telling – became infamous across the country, Target realized that a spike in a female customer's purchases of unscented body lotion would often precede her signing up to the in-store baby-shower registry. It had found a signal in the data. As women entered their second trimester and started to worry about stretch marks, their buying of moisturizer to keep their skin supple left a hint of what was to come. Scroll backwards further in time, and these same women would be popping into Target to stock up on various vitamins and supplements, like calcium and zinc. Scroll forwards in time and the data would even suggest when the baby was due – marked by the woman buying extra-big bags of cotton wool from the store.[6]

Expectant mothers are a retailer's dream. Lock in her loyalty while she's pregnant and there's a good chance she'll continue to use your products long after the birth of her child. After all, shopping habits are quick to form when a hungry screaming baby is demanding your attention during your weekly shop. Insights like this could be hugely valuable in giving Target a head start over other brands in attracting her business.

From there it was simple. Target ran an algorithm that would score its female customers on the likelihood they were pregnant. If that probability tipped past a certain threshold, the retailer would automatically send out a series of coupons to the woman in question, full of things she might find useful: nappies, lotions, baby wipes and so on.

So far, so uncontroversial. But then, around a year after the tool was first introduced, a father of a teenage girl stormed into a Target store in Minneapolis demanding to see the manager. His daughter had been sent some pregnancy coupons in the post and he was outraged that the retailer seemed to be normalizing teenage pregnancy. The manager of the store apologized profusely and called the man's home a few days later to reiterate the company's regret about the whole affair. But by then, according to a story in the *New York Times*, the father had an apology of his own to make.

'I had a talk with my daughter,' he told the manager. 'It turns out there's been some activities in my house I haven't been completely aware of. She's due in August.'

I don't know about you, but for me, an algorithm that will inform a parent that their daughter is pregnant before they've had a chance to learn about it in person takes a big step across the creepy line. But this embarrassment wasn't enough to persuade Target to scrap the tool altogether.

A Target executive explained: 'We found out that as long as a pregnant woman thinks she hasn't been spied on, she'll use the coupons. She just assumes that everyone else on her block got the same mailer for diapers and cribs. As long as we don't spook her, it works.'

So, Target still has a pregnancy predictor running behind the scenes – as most retailers do now. The only difference is that it will mix in the pregnancy-related coupons with other more generic items so that the customers don't notice they've been targeted. An advertisement for a crib might appear opposite some wine glasses. Or a coupon for baby clothes will run alongside an ad for some cologne.

Target is not alone in using these methods. Stories of what can be inferred from your data rarely hit the press, but the algorithms are out there, quietly hiding behind the corporate front lines. About a year ago, I got chatting to a chief data officer of a company that sells insurance. They had access to the full detail of people's shopping habits via a supermarket loyalty scheme. In their analysis, they'd discovered that home cooks were less likely to claim on their home insurance, and were therefore more profitable. It's a finding that makes good intuitive sense. There probably isn't much cross-over between the group of people who are willing to invest time, effort and money in creating an elaborate dish from scratch and the group who would let their children play football in the house.

But how did they know which shoppers were home cooks? Well, there were a few items in someone's basket that were linked to low claim rates. The most significant, he told me, the one that gives you away as a responsible, houseproud person more than any other, was fresh fennel.

If that's what you can infer from people's shopping habits in the physical world, just imagine what you might be able to infer if you had access to more data. Imagine how much you could learn about someone if you had a record of everything they did online.

The Wild West

Palantir Technologies is one of the most successful Silicon Valley start-ups of all time. It was founded in 2003 by Peter Thiel (of Pay-Pal fame), and at the last count was estimated to be worth a staggering $20 billion.[7] That's about the same market value as Twitter, although chances are you've never heard of it. And yet – trust me when I tell you – Palantir has most certainly heard of you.

Palantir is just one example of a new breed of companies known as data brokers, who buy and collect people's personal information and then resell it or share it for profit. There are plenty of others: Acxiom, Corelogic, Datalogix, eBureau – a swathe of huge companies you've probably never directly interacted with, that are none the less continually monitoring and analysing your behaviour.[8]

Every time you shop online, every time you sign up for a newsletter, or register on a website, or enquire about a new car, or fill out a warranty card, or buy a new home, or register to vote – every time you hand over any data at all – your information is being collected and sold to a data broker. Remember when you told an estate agent what kind of property you were looking for? Sold to a data broker. Or those details you once typed into an insurance comparison website? Sold to a data broker. In some cases, even your entire browser history can be bundled up and sold on.[9]

It's the broker's job to combine all of that data, cross-referencing the different pieces of information they've bought and acquired, and then create a single detailed file on you: a data profile of your digital shadow. In the most literal sense, within some of these brokers' databases, you could open up a digital file with your ID number on it (an ID you'll never be told) that contains traces of everything you've ever done. Your name, your date of birth, your religious affiliation, your vacation habits, your credit-card usage, your net worth, your weight, your height, your political affiliation, your gambling habits, your disabilities, the medication you use, whether you've had an abortion, whether your parents are divorced, whether you're easily addictable, whether you are a rape victim, your opinions on gun control, your projected sexual orientation, your real sexual orientation, and your gullibility. There are thousands and thousands of details within thousands and thousands of categories and files stored on hidden servers somewhere, for virtually every single one of us.[10]

Like Target's pregnancy predictions, much of this data is inferred. A subscription to *Wired* magazine might imply that you're interested in technology; a firearms licence might imply that you're interested in hunting. All along the way, the brokers are using clever, but simple, algorithms to enrich their data. It's exactly what the supermarkets were doing, but on a massive scale.

And there are plenty of benefits to be had. Data brokers use their understanding of who we are to prevent fraudsters from impersonating unsuspecting consumers. Likewise, knowing our likes and dislikes means that the adverts we're served as we wander around the internet are as relevant to our interests and needs as possible. That almost certainly makes for a more pleasant experience than being hit with mass market adverts for injury lawyers or PPI claims day after day. Plus, because the messages can be directly targeted on the right consumers, it means advertising is cheaper overall, so small businesses with great products can reach new audiences, something that's good for everyone.

But, as I'm sure you're already thinking, there's also an array of problems that arise once you start distilling who we are as people down into a series of categories. I'll get on to that in a moment, but first I think it's worth briefly explaining the invisible process behind how an online advert reaches you when you're clicking around on the internet, and the role that a data broker plays in the process.

So, let's imagine I own a luxury travel company, imaginatively called Fry's. Over the years, I have been getting people to register their interest on my website and now have a list of their email addresses. If I wanted to find out more about my users – like what kind of holidays they were interested in – I could send off my list of users' emails to a data broker, who would look up the names in their system, and return my list with the relevant data attached. Sort of like adding an extra column on to a spreadsheet. Now when you visit my Fry's website, I can see that you have a particular penchant for tropical islands and so serve you up an advert for a Hawaii getaway.

That's option one. In option two, let's imagine that Fry's has a little extra space on its website that we're willing to sell to other advertisers. Again, I contact a data broker and give them the information I have on my users. The broker looks for other companies who want to place adverts. And, for the sake of the story, let's imagine that a company selling sun cream is keen. To persuade them that Fry's has the audience the sun-cream seller would want to target, the broker could show them some inferred characteristics of Fry's users: perhaps the percentage of people with red hair, that kind of thing. Or the sun-cream seller could hand over a list of its own users' email addresses and the broker could work out exactly how much crossover there was between the audiences. If the sun-cream seller agrees, the advert appears on Fry's website – and the broker and I both get paid.

So far, these methods don't go much beyond the techniques that marketers have always used to target customers. But it's option three where, for me, things start to get a little bit creepy. This time, Fry's is looking for some new customers. I want to target men and women over 65 who like tropical islands and have large disposable incomes, in the hope that they'll want to go on one of our luxurious new Caribbean cruises. Off I go to a data broker who will look through their database and find me a list of people who match my description.

So, let's imagine you are on that list. The broker will never share your name with Fry's. But they will work out which other websites you regularly use. Chances are, the broker will also have a relationship with one of your favourites. Maybe a social media site, or a news website, something along those lines. As soon as you unsuspectingly log into your favourite website, the broker will get a ping to alert them to the fact that you're there. Virtually instantaneously, the broker will respond by placing a little tiny flag – known as a cookie – on your computer. This cookie* acts like a signal to all kinds of other websites around the internet, saying that you are someone who should be served up an advert for Fry's Caribbean cruises. Whether you want them or not, wherever you go on the internet, those adverts will follow you.

And here we stumble on the first problem. What if you don't want to see the advert? Sure, being bombarded with images of Caribbean cruises might be little more than a minor inconvenience, but there

---

* Adverts aren't the only reason for cookies. They're also used by websites to see if you're logged in or not (to know if it's safe to send through any sensitive information) and to see if you're a returning visitor to a page (to trigger a price hike on an airline website, for instance, or email you a discount code on an online clothing store).

are other adverts which can have a much more profound impact on a person.

When Heidi Waterhouse lost a much-wanted pregnancy,[11] she unsubscribed from all the weekly emails updating her on her baby's growth, telling her which fruit the foetus now matched in size. She unsubscribed from all the mailing lists and wish lists she had signed up to in eager anticipation of the birth. But, as she told an audience of developers at a conference in 2018, there was no power on earth that could unsubscribe her from the pregnancy adverts that followed her around the internet. This digital shadow of a pregnancy continued to circulate alone, without the mother or the baby. 'Nobody who built that system thought of that consequence,' she explained.

It's a system which, thanks to either thoughtless omission or deliberate design, has the potential to be exploitative. Payday lenders can use it to directly target people with bad credit ratings; betting adverts can be directed to people who frequent gambling websites. And there are concerns about this kind of data profiling being used against people, too: motorbike enthusiasts being deemed to have a risky hobby, or people who eat sugar-free sweets being flagged as diabetic and turned down for insurance as a result. A study from 2015 demonstrated that Google was serving far fewer ads for high-paying executive jobs to women who were surfing the web than to men.[12] And, after one African American Harvard professor learned that Googling her own name returned adverts targeted on people with a criminal record (and as a result was forced to prove to a potential employer that she'd never been in trouble with the police), she began researching the adverts delivered to different ethnic groups. She discovered that searches for 'black-sounding names' were disproportionately likely to be linked to adverts containing the word 'arrest' (e.g. 'Have you been arrested?') than those with 'white-sounding names'.[13]

These methods aren't confined to data brokers. There's very little difference between how they work and how Google, Facebook, Instagram and Twitter operate. These internet giants don't make money by having users, so their business models are based on the idea of micro-targeting. They are gigantic engines for delivering adverts, making money by having their millions of users actively engaged on their websites, clicking around, reading sponsored posts, watching sponsored videos, looking at sponsored photos. In whatever corner of the internet you use, hiding in the background, these algorithms are trading on information you didn't know they had and never willingly offered. They have made your most personal, private secrets into a commodity.

Unfortunately, in many countries, the law doesn't do much to protect you. Data brokers are largely unregulated and – particularly in America – opportunities to curb their power have repeatedly been passed over by government. In March 2017, for instance, the US Senate voted to eliminate rules that would have prevented data brokers from selling your internet browser history without your explicit consent. Those rules had previously been approved in October 2016 by the Federal Communications Commission; but, after the change in government at the end of that year, they were opposed by the FCC's new Republican majority and Republicans in Congress.[14]

So what does all this mean for your privacy? Well, let me tell you about an investigation led by German journalist Svea Eckert and data scientist Andreas Dewes that should give you a clear idea.[15]

Eckert and her team set up a fake data broker and used it to buy the anonymous browsing data of 3 million German citizens. (Getting hold of people's internet histories was easy. Plenty of companies had an abundance of that kind of data for sale on British or US customers – the only challenge was finding data focused on Germany.) The data itself had been gathered by a Google Chrome plugin that users had willingly downloaded, completely unaware that it was spying on them in the process.*

In total, it amounted to a gigantic list of URLs. A record of everything those people had looked at online over the course of a month. Every search, every page, every click. All legally put up for sale.

For Eckert and her colleagues, the only problem was that the browser data was anonymous. Good news for all the people whose histories had been sold. Right? Should save their blushes. Wrong. As the team explained in a presentation at DEFCON in 2017, de-anonymizing huge databases of browser history was spectacularly easy.

Here's how it worked. Sometimes there were direct clues to the person's identity in the URLs themselves. Like anyone who visited Xing.com, the German equivalent of LinkedIn. If you click on your profile picture on the Xing website, you are sent through to a page with an address that will be something like the following:

www.xing.com/profile/Hannah_Fry?sc_omxb_p

Instantly, the name gives you away, while the text after the username signifies that the user is logged in and viewing their own profile, so the team could be certain that the individual was looking at their own page. It was a similar story with Twitter. Anyone checking their own Twitter analytics page was revealing themselves to the team in the process. For those without an instant identifier in their data, the team had another trick up their sleeve. Anyone who posted a link online – perhaps by tweeting about a website, or sharing their public playlist on YouTube – essentially, anyone who left a public trace of their data shadow attached to their real name, was inadvertently unmasking themselves in the process. The team used

* That plugin, ironically called 'The Web of Trust', set out all this information clearly in black and white as part of the terms and conditions.

a simple algorithm to cross-reference the public and anonymized personas,[16] filtering their list of URLs to find someone in the dataset who had visited the same websites at the same times and dates that the links were posted online. Eventually, they had the full names of virtually everyone in the dataset, and full access to a month's worth of complete browsing history for millions of Germans as a result.

Among those 3 million people were several high-profile individuals. They included a politician who had been searching for medication online. A police officer who had copied and pasted a sensitive case document into Google Translate, all the details of which then appeared in the URL and were visible to the researchers. And a judge, whose browsing history showed a daily visit to one rather specific area of the internet. Here is a small selection of the websites he visited during one eight-minute period in August 2016:

18.22: http://www.tubegalore.com/video/amature-pov–ex-wife-in-leather-pants-gets-creampie42945.html

18.23: http://www.xxkingtube.com/video/pov_wifey_on_sex_stool_with_beaded_thong_gets_creampie_4814.html

18.24: http://de.xhamster.com/movies/924590/office_lady_in_pants_rubbing_riding_best_of_anlife.html

18.27: http://www.tubegalore.com/young_tube/5762–1/page0

18.30: http://www.keezmovies.com/video/sexy-dominatrix-milks-him-dry-1007114?utm_sources

In among these daily browsing sessions, the judge was also regularly searching for baby names, strollers and maternity hospitals online. The team concluded that his partner was expecting a baby at the time.

Now, let's be clear here: this judge wasn't doing anything illegal. Many – myself included – would argue that he wasn't doing anything wrong at all. But this material would none the less be useful in the hands of someone who wanted to blackmail him or embarrass his family.

And that is where we start to stray very far over the creepy line. When private, sensitive information about you, gathered without your knowledge, is then used to manipulate you. Which, of course, is precisely what happened with the British political consulting firm Cambridge Analytica.

Cambridge Analytica

You probably know most of the story by now.

Since the 1980s, psychologists have been using a system of five characteristics to quantify an individual's personality. You get a score on each of the following traits: openness to experience, conscientiousness, extraversion, agreeableness and neuroticism. Collectively, they offer a standard and useful way to describe what kind of a person you are.

Back in 2012, a year before Cambridge Analytica came on the scene, a group of scientists from the University of Cambridge and Stanford University began looking for a link between the five personality traits and the pages people 'liked' on Facebook.[17] They built a Facebook quiz with this purpose in mind, allowing users to take real psychometric tests, while hoping to find a connection between a person's true character and their online persona. People who downloaded their quiz knowingly handed over data on both: the history of their Likes on Facebook and, through a series of questions, their true personality scores.

It's easy to imagine how Likes and personality might be related. As the team pointed out in the paper they published the following year,[18] people who like Salvador Dalí, meditation or TED talks are almost certainly going to score highly on openness to experience. Meanwhile, people who like partying, dancing and Snooki from the TV series *Jersey Shore* tend to be a bit more extraverted. The research was a success. With a connection established, the team built an algorithm that could infer someone's personality from their Facebook Likes alone.

By the time their second study appeared in 2014,[19] the research team were claiming that if you could collect 300 Likes from someone's Facebook profile, the algorithm would be able to judge their character more accurately than their spouse could.

Fast-forward to today, and the academic research group – the Psychometrics Centre at Cambridge University – have extended their algorithm to make personality predictions from your Twitter feed too. They have a website, open to anyone, where you can try it for yourself. Since my Twitter profile is open to the public anyway, I thought I'd try out the researchers' predictions myself, so uploaded my Twitter history and filled out a traditional questionnaire-based personality study to compare. The algorithm managed to assess me accurately on three of the five traits. Although, as it turns out, according to the traditional personality study I am much more extraverted and neurotic than my Twitter profile makes it seem.*

All this work was motivated by how it could be used in advertising. So, by 2017,[20] the same team of academics had moved on to experimenting with sending out adverts tailored to an individual's personality traits. Using the Facebook platform, the team served up adverts for a beauty product to extraverts using the slogan 'Dance like no one's watching (but they totally are)', while introverts saw an image of a girl smiling and standing in front of the mirror with the phrase 'Beauty doesn't have to shout.'

In a parallel experiment, targets high in openness-to-experience were shown adverts for crossword puzzles using an image with the text: 'Aristoteles? The Seychelles? Unleash your creativity and

---

* That particular combination seems to imply that I'd post more stuff if I didn't get so worried about how it'd go down.

challenge your imagination with an unlimited number of crossword puzzles!' The same puzzles were advertised to people low in openness, but using instead the wording: 'Settle in with an all-time favorite! The crossword puzzle that has challenged players for generations.' Overall, the team claimed that matching adverts to a person's character led to 40 per cent more clicks and up to 50 per cent more purchases than using generic, unpersonalized ads. For an advertiser, that's pretty impressive.

All the while, as the academics were publishing their work, others were implementing their methods. Among them, so it is alleged, was Cambridge Analytica during their work for Trump's election campaign.

Now, let's backtrack slightly. There is little doubt that Cambridge Analytica were using the same techniques as my imaginary luxury travel agency Fry's. Their approach was to identify small groups of people who they believed to be persuadable and target them directly, rather than send out blanket advertising. As an example they discovered that there was a large degree of overlap between people who bought good, American-made Ford motor cars and people who were registered as Republican party supporters. So they then set about finding people who had a preference for Ford, but weren't known Republican voters, to see if they could sway their opinions using all-American adverts that tapped into that patriotic emotion. In some sense, this is no different from a candidate identifying a particular neighbourhood of swing voters and going door-to-door to persuade them one by one. And online, it's no different from what Obama and Clinton were doing during their campaigns. Every major political party in the Western world uses extensive analysis and micro-targeting of voters.

But, if the undercover footage recorded by *Channel Four News* is to be believed, Cambridge Analytica were also using personality profiles of the electorate to deliver emotionally charged political messages – for example, finding single mothers who score highly on neuroticism and preying on their fear of being attacked in their own home to persuade them into supporting a pro-gun-lobby message. Commercial advertisers have certainly used these techniques extensively, and other political campaigns probably have, too.

But on top of all that, Cambridge Analytica are accused of creating adverts and dressing them up as journalism. According to one whistleblower's testimony to the *Guardian*, one of the most effective ads during the campaign was an interactive graphic titled '10 inconvenient truths about the Clinton Foundation'.[21] Another whistleblower went further and claimed that the 'articles' planted by Cambridge Analytica were often based on demonstrable falsehoods.[22]

Let's assume, for the sake of argument, that all the above is true: Cambridge Analytica served up manipulative fake news stories on Facebook to people based on their psychological profiles. The question is, did it work?

Micro-manipulation

There is an asymmetry in how we view the power of targeted political adverts. We like to think of ourselves as independently minded and immune to manipulation, and yet imagine others – particularly those of a different political persuasion – as being fantastically gullible. The reality is probably something in between.

We do know that the posts we see on Facebook have the power to alter our emotions. A controversial experiment run by Facebook employees in 2013 manipulated the news feeds of 689,003 users without their knowledge (or consent) in an attempt to control their emotions and influence their moods.[23] The experimenters suppressed any friends' posts that contained positive words, and then did the same with those containing negative words, and watched to see how the unsuspecting subjects would react in each case. Users who saw less negative content in their feeds went on to post more positive stuff themselves. Meanwhile, those who had positive posts hidden from their timeline went on to use more negative words themselves. Conclusion: we may think we're immune to emotional manipulation, but we're probably not.

We also know from the Epstein experiment described in the 'Power' chapter that just the ordering of pages on a search engine can be enough to tip undecided voters into favouring one candidate over another. We know, too, from the work done by the very academics whose algorithms Cambridge Analytica repurposed, that adverts are more effective if they target personality traits.

Put together, all this does build a strong argument to suggest that these methods can have an impact on how people vote, just as they do on how people spend their money. But – and it's quite a big but – there's something else you need to know before you make your mind up.

All of the above is true, but the actual effects are tiny. In the Facebook experiment, users were indeed more likely to post positive messages if they were shielded from negative news. But the difference amounted to less than one-tenth of one percentage point.

Likewise, in the targeted adverts example, the makeup sold to introverts was more successful if it took into account the person's character, but the difference it made was minuscule. A generic advert got 31 people in 1,000 to click on it. The targeted ad managed 35 in 1,000. Even that figure of 50 per cent improvement that I cited on page 41, which is boldly emblazoned across the top of the academic paper, is actually referring to an increase from 11 clicks in 1,000 to 16.

The methods can work, yes. But the advertisers aren't injecting their messages straight into the minds of a passive audience. We're not sitting ducks. We're much better at ignoring advertising or putting our own spin on interpreting propaganda than the people sending those messages would like us to be. In the end, even with the best, most deviously micro-profiled campaigns, only a small amount of influence will leak through to the target.

And yet, potentially, in an election those tiny slivers of influence might be all you need to swing the balance. In a population of tens or hundreds of millions, those one-in-a-thousand switches can quickly add up. And when you remember that, as Jamie Bartlett pointed out in a piece for the *Spectator*, Trump won Pennsylvania by 44,000 votes out of six million cast, Wisconsin by 22,000, and Michigan by 11,000, perhaps margins of less than 1 per cent might be all you need.[24]

The fact is, it's impossible to tell just how much of an effect all this had in the US presidential election. Even if we had access to all of the facts, we can't look back through time and untangle the sticky web of cause and effect to pinpoint a single reason for anyone's voting decisions. What has gone has gone. What matters now is where we go in the future.

Rate me

It's important to remember that we've all benefited from this model of the internet. All around the world, people have free and easy access to instant global communication networks, the wealth of human knowledge at their fingertips, up-to-the-minute information from across the earth, and unlimited usage of the most remarkable software and technology, built by private companies, paid for by adverts. That was the deal that we made. Free technology in return for your data and the ability to use it to influence and profit from you. The best and worst of capitalism in one simple swap.

We might decide we're happy with that deal. And that's perfectly fine. But if we do, it's important to be aware of the dangers of collecting this data in the first place. We need to consider where these datasets could lead – even beyond the issues of privacy and the potential to undermine democracy (as if they weren't bad enough). There is another twist in this dystopian tale. An application for these rich, interconnected datasets that belongs in the popular Netflix show *Black Mirror*, but exists in reality. It's known as Sesame Credit, a citizen scoring system used by the Chinese government.

Imagine every piece of information that a data broker might have on you collapsed down into a single score. Everything goes into it. Your credit history, your mobile phone number, your address – the usual stuff. But all your day-to-day behaviour, too. Your social media posts, the data from your ride-hailing app, even records from your online matchmaking service. The result is a single number between 350 and 950 points.

Sesame Credit doesn't disclose the details of its 'complex' scoring algorithm. But Li Yingyun, the company's technology director, did share some examples of what might be inferred from its results in an interview with the Beijing-based Caixin Media. 'Someone who plays video games for ten hours a day, for example, would be considered an idle person. Someone who frequently buys diapers would be considered as probably a parent, who on balance is more likely to have a sense of responsibility.'[25]

If you're Chinese, these scores matter. If your rating is over 600 points, you can take out a special credit card. Above 666 and you'll be rewarded with a higher credit limit. Those with scores above 650 can hire a car without a deposit and use a VIP lane at Beijing airport. Anyone over 750 can apply for a fast-tracked visa to Europe.[26]

It's all fun and games now while the scheme is voluntary. But when the citizen scoring system becomes mandatory in 2020, people with low scores stand to feel the repercussions in every aspect of their lives. The government's own document on the system outlines examples of punishments that could be meted out to anyone deemed disobedient: 'Restrictions on leaving the borders, restrictions on the purchase of ... property, travelling on aircraft, on tourism and holidays or staying in star-ranked hotels.' It also warns that in the case of 'gravely trust breaking subjects' it will 'guide commercial banks ... to limit their provision of loans, sales insurance and other such services.'[27] Loyalty is praised. Breaking trust is punished. As Rogier Creemers, an academic specializing in Chinese law and governance at the Van Vollenhoven Institute at Leiden University, puts it: 'The best way to understand it is as a sort of bastard love child of a loyalty scheme.'[28]

I don't have much comfort to offer in the case of Sesame Credit, but I don't want to fill you completely with doom and gloom, either. There are glimmers of hope elsewhere. However grim the journey ahead appears, there are signs that the tide is slowly turning. Many in the data science community have known about and objected to the exploitation of people's information for profit for quite some time. But until the furore over Cambridge Analytica these issues hadn't drawn sustained, international front-page attention. When that scandal broke in early 2018 the general public saw for the first time how algorithms are silently harvesting their data, and acknowledged that, without oversight or regulation, it could have dramatic repercussions.

And regulation is coming. If you live in the EU, there has recently been a new piece of legislation called GDPR – General Data Protection Regulation – that should make much of what data brokers are doing illegal. In theory, they will no longer be allowed to store your data without an explicit purpose. They won't be able to infer information about you without your consent. And they won't be able to get your permission to collect your data for one reason, and then secretly use it for another. That doesn't necessarily mean the end of these kinds of practices, however. For one thing, we often don't pay attention to the T&Cs when we're clicking around online, so we may find ourselves consenting without realizing. For another, the identification of illegal practices and enforcement of regulations remains tricky in a world where most data analysis and transfer happens in the shadows. We'll have to wait and see how this unfolds.

Europeans are the lucky ones, but there are those pushing for regulation in America, too. The Federal Trade Commission published a report condemning the murky practices of data brokers back in 2014, and since then has been actively pushing for more consumer rights. Apple has now built 'intelligent tracking prevention' into the Safari browser. Firefox has done the same. Facebook is severing ties with its data brokers. Argentina and Brazil, South Korea and many more countries have all pushed through GDPR-like legislation. Europe might be ahead of the curve, but there is a global trend that is heading in the right direction.

If data is the new gold, then we've been living in the Wild West. But I'm optimistic that – for many of us – the worst will soon be behind us.

Still, we do well to remember that there's no such thing as a free lunch. As the law catches up and the battle between corporate profits and social good plays out, we need to be careful not to be lulled into a false sense of privacy. Whenever we use an algorithm – especially a free one – we need to ask ourselves about the hidden incentives. Why is this app giving me all this stuff for free? What is this algorithm really doing? Is this a trade I'm comfortable with? Would I be better off without it?

That is a lesson that applies well beyond the virtual realm, because the reach of these kinds of calculations now extends into virtually every aspect of society. Data and algorithms don't just have the power to predict our shopping habits. They also have the power to rob someone of their freedom.

Justice

It's not unusual to find good-natured revellers drinking on a summer Sunday evening in the streets of Brixton, where our next story begins. Brixton, in south London, has a reputation as a good place to go for a night out; on this particular evening, a music festival had just finished and the area was filled with people merrily making their way home, or carrying on the party. But at 11.30 p.m. the mood shifted. A fight broke out on a local council estate, and when police failed to contain the trouble, it quickly spilled into the centre of Brixton, where hundreds of young people joined in.

This was August 2011. The night before, on the other side of the city, an initially peaceful protest over the shooting by police of a young Tottenham man named Mark Duggan had turned violent. Now, for the second night in a row, areas of the city were descending into chaos – and this time, the atmosphere was different. What had begun as a local demonstration was now a widespread breakdown in law and order and a looting free-for-all.

Just as the rioting took hold, Nicholas Robinson, a 23-year-old electrical engineering student who had spent the weekend at his girlfriend's, headed home, taking his usual short walk through Brixton.[1] By now, the familiar streets were practically unrecognizable. Cars had been upturned, windows had been smashed, fires had been started, and all along the street shops had been broken into.[2] Police had been desperately trying to calm the situation, but were powerless to stop the cars and scooters pulling up alongside the smashed shop fronts and loading up with stolen clothes, shoes, laptops and TVs. Brixton was completely out of control.

A few streets away from an electrical store that was being thoroughly ransacked, Nicholas Robinson walked past his local supermarket. Like almost every other shop, it was a wreck: the glass windows and doors had been broken, and the shelves inside were strewn with the mess from the looters. Streams of rioters were running past holding on to their brand-new laptops, unchallenged by police officers. Amid the chaos, feeling thirsty, Nicholas walked into the store and helped himself to a £3.50 case of bottled water. Just as he rounded the corner to leave, the police entered the supermarket. Nicholas immediately realized what he had done, dropped the case and tried to run.[3]

As Monday night rolled in, the country braced itself for further riots. Sure enough, that night looters took to the streets again.[4] Among them was 18-year-old Richard Johnson. Intrigued by what he'd seen on the news, he grabbed a (distinctly un-summery) balaclava, jumped into a car and made his way to the local shopping centre. With his face concealed, Richard ran into the town's gaming store, grabbed a haul of computer games and returned to the car.[5] Unfortunately for Richard, he had parked in full view of a CCTV camera. The registration plate made it easy for police to track him down, and the recorded evidence made it easy to bring a case against him.[6]

Both Richard Johnson and Nicholas Robinson were arrested for their actions in the riots of 2011. Both were charged with burglary. Both stood before judges. Both pleaded guilty. But that's where the similarities in their cases end.

Nicholas Robinson was first to be called to the dock, appearing before a judge at Camberwell Magistrates' Court less than a week after the incident. Despite the low value of the bottled water he had stolen, despite his lack of a criminal record, his being in full-time education, and his telling the court he was ashamed of himself, the judge said his actions had contributed to the atmosphere of lawlessness in Brixton that night. And so, to gasps from his family in the public gallery, Nicholas Robinson was sentenced to six months in prison.[7]

Johnson's case appeared in front of a judge in January 2012. Although he'd gone out wearing an item of clothing designed to hide his identity with the deliberate intention of looting, and although he too had played a role in aggravating public disorder, Johnson escaped jail entirely. He was given a suspended sentence and ordered to perform two hundred hours of unpaid work.[8]

The consistency conundrum

The judicial system knows it's not perfect, but it doesn't try to be. Judging guilt and assigning punishment isn't an exact science, and there's no way a judge can guarantee precision. That's why phrases such as 'reasonable doubt' and 'substantial grounds' are so fundamental to the legal vocabulary, and why appeals are such an important part of the process; the system accepts that absolute certainty is unachievable.

Even so, discrepancies in the treatment of some defendants – like Nicholas Robinson and Richard Johnson – do seem unjust. There are too many factors involved ever to say for certain that a difference in sentence is 'unfair', but within reason, you would hope that judges were broadly consistent in the way they made decisions. If you and your imaginary twin committed an identical crime, for instance, you would hope that a court would give you both the same sentence. But would it?

In the 1970s, a group of American researchers tried to answer a version of this question.[9] Rather than using twin criminals (which is practically difficult and ethically undesirable) they created a series of hypothetical cases and independently asked 47 Virginia

State district court judges how they would deal with each. Here's an example from the study for you to try. How would you handle the following case?

An 18-year-old female defendant was apprehended for possession of marijuana, arrested with her boyfriend and seven other acquaintances. There was evidence of a substantial amount of smoked and un-smoked marijuana found, but no marijuana was discovered directly on her person. She had no previous criminal record, was a good student from a middle-class home and was neither rebellious nor apologetic for her actions.

The differences in judgments were dramatic. Of the 47 judges, 29 decreed the defendant not guilty and 18 declared her guilty. Of those who opted for a guilty verdict, eight recommended probation, four thought a fine was the best way to go, three would issue both probation and a fine, and three judges were in favour of sending the defendant to prison.

So, on the basis of identical evidence in identical cases, a defendant could expect to walk away scot-free or be sent straight to jail, depending entirely on which judge they were lucky (or unlucky) enough to find themselves in front of.

That's quite a blow for anyone holding on to hopes of courtroom consistency. But it gets worse. Because, not only do the judges disagree with each other, they're also prone to contradicting *their own decisions*.

In a more recent study, 81 UK judges were asked whether they'd award bail to a number of imaginary defendants.[10] Each hypothetical case had its own imaginary back-story and imaginary criminal history. Just like their counterparts in the Virginia study, the British judges failed to agree unanimously on a single one of the 41 cases presented to them.[11] But this time, in among the 41 hypothetical cases given to every judge were seven that appeared twice – with the names of the defendants changed on their second appearance so the judge wouldn't notice they'd been duplicated. It was a sneaky trick, but a revealing one. Most judges didn't manage to make the same decision on the same case when seeing it for the second time. Astonishingly, some judges did no better at matching their own answers than if they were, quite literally, awarding bail at random.[12]

Numerous other studies have come to the same conclusion: whenever judges have the freedom to assess cases for themselves, there will be massive inconsistencies. Allowing judges room for discretion means allowing there to be an element of luck in the system.

There is a simple solution, of course. An easy way to make sure that judges are consistent is to take away their ability to exercise discretion. If every person charged with the same offence were dealt with in exactly the same way, precision could be guaranteed – at least in matters of bail and sentencing. And indeed, some countries have taken this route. There are prescriptive sentencing systems in operation at the federal level in the United States and in parts of Australia.[13] But this kind of consistency comes at a price. For all that you gain in precision, you lose in another kind of fairness.

To illustrate, imagine two defendants, both charged with stealing from a supermarket. One is a relatively comfortable career criminal stealing through choice; the other is recently unemployed and struggling to make ends meet, stealing to feed their family and full of remorse for their actions. By removing the freedom to take these mitigating factors into account, strict guidelines that treat all those charged with the same crime in the same way can end up being unnecessarily harsh on some, and mean throwing away the chance to rehabilitate some criminals.

This is quite a conundrum. However you set up the system for your judges, you have to find a tricky balance between offering individualized justice and ensuring consistency. Most countries have tried to solve the dilemma by settling on a system that falls somewhere between the prescriptive extreme of US federal law and one based almost entirely on judges' discretion – like that used in Scotland.[14] Across the Western world, sentencing guidelines tend to lay down a maximum sentence (as in Ireland) or a minimum sentence (as in Canada) or both (as in England and Wales),[15] and allow judges latitude to adjust the sentence up or down between those limits.

No system is perfect. There is always a muddle of competing unfairnesses, a chaos of opposing injustices. But in all this conflict and complexity an algorithm has a chance of coming into its own. Because – remarkably – with an algorithm as part of the process, both consistency and individualized justice can be guaranteed. No one needs to choose between them.

The justice equation

Algorithms can't decide guilt. They can't weigh up arguments from the defence and prosecution, or analyse evidence, or decide whether a defendant is truly remorseful. So don't expect them to replace judges any time soon. What an algorithm can do, however, incredible as it might seem, is use data on an individual to calculate their risk of re-offending. And, since many judges' decisions are based on the likelihood that an offender will return to crime, that turns out to be a rather useful capacity to have.

Data and algorithms have been used in the judicial system for almost a century, the first examples dating back to 1920s America. At the time, under the US system, convicted criminals would be sentenced to a standard maximum term and then become eligible for parole* after a period of time had elapsed. Tens of thousands

---

* Fun fact: 'parole' comes from the French *parole*, meaning 'voice, spoken words'. It originated in its current form in the 1700s, when prisoners would be released if they gave their word that they would not return to crime: https://www.etymonline.com/word/parole.

of prisoners were granted early release on this basis. Some were successfully rehabilitated, others were not. But collectively they presented the perfect setting for a natural experiment: could you predict whether an inmate would violate their parole?

Enter Ernest W. Burgess, a Canadian sociologist at the University of Chicago with a thirst for prediction. Burgess was a big proponent of quantifying social phenomena. Over the course of his career he tried to forecast everything from the effects of retirement to marital success, and in 1928 he became the first person to successfully build a tool to predict the risk of criminal behaviour based on measurement rather than intuition.

Using all kinds of data from three thousand inmates in three Illinois prisons, Burgess identified 21 factors he deemed to be 'possibly significant' in determining the chances of whether someone would violate the terms of their parole. These included the type of offence, the months served in prison and the inmate's social type, which – with the delicacy one would expect from an early-twentieth-century social scientist – he split into categories including 'hobo', 'drunkard', 'ne'er do-well', 'farm boy' and 'immigrant'.[16]

Burgess gave each inmate a score between zero and one on each of the 21 factors. The men who got high scores (between 16 and 21) he deemed least likely to re-offend; those who scored low (four or less) he judged likely to violate their terms of release.

When all the inmates were eventually granted their release, and so were free to violate the terms of their parole if they chose to, Burgess had a chance to check how good his predictions were. From such a basic analysis, he managed to be remarkably accurate. Ninety-eight per cent of his low-risk group made a clean pass through their parole, while two-thirds of his high-risk group did not.[17] Even crude statistical models, it turned out, could make better forecasts than the experts.

But his work had its critics. Sceptical onlookers questioned how much the factors which reliably predicted parole success in one place at one time could apply elsewhere. (They had a point: I'm not sure the category 'farm boy' would be much help in predicting recidivism among modern inner-city criminals.) Other scholars criticized Burgess for just making use of whatever information was on hand, without investigating if it was relevant.[18] There were also questions about the way he scored the inmates: after all, his method was little more than opinion written in equations. None the less, its forecasting power was impressive enough that by 1935 the Burgess method had made its way into Illinois prisons, to support parole boards in making their decisions.[19] And by the turn of the century mathematical descendants of Burgess's method were being used all around the world.[20]

Fast-forward to the modern day, and the state-of-the-art risk-assessment algorithms used by courtrooms are far more sophisticated than the rudimentary tools designed by Burgess. They are not only found assisting parole decisions, but are used to help match intervention programmes to prisoners, to decide who should be awarded bail, and, more recently, to support judges in their sentencing decisions. The fundamental principle is the same as it always was: in go the facts about the defendant – age, criminal history, seriousness of the crime and so on – and out comes a prediction of how risky it would be to let them loose.

So, how do they work? Well, broadly speaking, the best-performing contemporary algorithms use a technique known as *random forests*, which – at its heart – has a fantastically simple idea. The humble decision tree.

Ask the audience

You might well be familiar with decision trees from your schooldays. They're popular with maths teachers as a way to structure observations, like coin flips or dice rolls. Once built, a decision tree can be used as a flowchart: taking a set of circumstances and assessing step by step what to do, or, in this case, what will happen.

Imagine you're trying to decide whether to award bail to a particular individual. As with parole, this decision is based on a straightforward calculation. Guilt is irrelevant. You only need to make a prediction: will the defendant violate the terms of their bail agreement, if granted leave from jail?

To help with your prediction, you have data from a handful of previous offenders, some who fled or went on to re-offend while on bail, some who didn't. Using the data, you could imagine constructing a simple decision tree by hand, like the one below, using the characteristics of each offender to build a flowchart. Once built, the decision tree can forecast how the new offender might behave. Simply follow the relevant branches according to the characteristics of the offender until you get to a prediction. Just as long as they fit the pattern of everyone who has gone before, the prediction will be right.

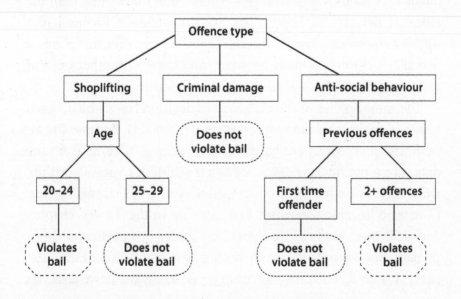

But this is where decision trees of the kind we made in school start to fall down. Because, of course, not everyone *does* follow the pattern of those who went before. On its own, this tree is going to get a lot of forecasts wrong. And not just because we're starting with a simple example. Even with an enormous dataset of previous cases and an enormously complicated flowchart to match, using a single tree may only ever be slightly better than random guessing.

And yet, if you build more than one tree – everything can change. Rather than using all the data at once, there is a way to divide and conquer. In what is known as an *ensemble*, you first build thousands of smaller trees from random subsections of the data. Then, when presented with a new defendant, you simply ask every tree to vote on whether it thinks awarding bail is a good idea or not. The trees may not all agree, and on their own they might still make weak predictions, but just by taking the average of all their answers, you can dramatically improve the precision of your predictions.

It's a bit like asking the audience in *Who Wants To Be A Millionaire?* A room full of strangers will be right more often than the cleverest person you know. (The 'ask the audience' lifeline had a 91 per cent success rate compared to just 65 per cent for 'phone a friend'.?)[21] The errors made by many can cancel each other out and result in a crowd that's wiser than the individual.

The same applies to the big group of decision trees which, taken together, make up a random forest (pun intended). Because the algorithm's predictions are based on the patterns it learns from the data, a random forest is described as a machine-learning algorithm, which comes under the broader umbrella of artificial intelligence. (The term 'machine learning' first came up in the 'Power' chapter, and we'll meet many more algorithms under this particular canopy later, but for now it's worth noting how grand that description makes it sound, when the algorithm is essentially the flowcharts you used to draw at school, wrapped up in a bit of mathematical manipulation.) Random forests have proved themselves to be incredibly useful in a whole host of real-world applications. They're used by Netflix to help predict what you'd like to watch based on past preferences;[22] by Airbnb to detect fraudulent accounts;[23] and in healthcare for disease diagnosis (more on that in the following chapter).

When used to assess offenders, they can claim two huge advantages over their human counterparts. First, the algorithm will always give exactly the same answer when presented with the same set of circumstances. Consistency comes guaranteed, but not at the price of individualized justice. And there is another key advantage: the algorithm also makes much better predictions.

Human vs machine

In 2017, a group of researchers set out to discover just how well a machine's predictions stacked up against the decisions of a bunch of human judges.[24]

To help them in their mission, the team had access to the records of every person arrested in New York City over a five-year period between 2008 and 2013. During that time, three-quarters of a million people were subject to a bail hearing, which meant easily enough data to test an algorithm on a head-to-head basis with a human judge.

An algorithm hadn't been used by the New York judicial system during these cases, but looking retrospectively, the researchers got to work building lots of decision trees to see how well one could have predicted the defendants' risk of breaking bail conditions at the time. In went the data on an offender: their rap sheet, the crime they'd just committed and so on. Out came a probability of whether or not that defendant would go on to violate the terms of their bail.

In the real data, 408,283 defendants were released before they faced trial. Any one of those was free to flee or commit other crimes, which means we can use the benefit of hindsight to test how accurate the algorithm's predictions and the humans' decisions were. We know exactly who failed to appear in court later (15.2 per cent) and who was re-arrested for another crime while on bail (25.8 per cent).

Unfortunately for the science, any defendant deemed high risk by a judge would have been denied bail at the time – and hence, on those cases, there was no opportunity to prove the judges' assessment right or wrong. That makes things a little complicated. It means there's no way to come up with a cold, hard number that captures how accurate the judges were overall. And without a 'ground truth' for how those defendants would have behaved, you can't state an overall accuracy for the algorithm either. Instead, you have to make an educated guess on what the jailed defendants would have done if released,[25] and make your comparisons of human versus machine in a bit more of a roundabout way.

One thing is for sure, though: the judges and the machine didn't agree on their predictions. The researchers showed that many of the defendants flagged by the algorithm as real bad guys were treated by the judges as though they were low risk. In fact, almost half of the defendants the algorithm flagged as the riskiest group were given bail by the judges.

But who was right? The data showed that the group the algorithm was worried about did indeed pose a risk. Just over 56 per cent of them failed to show up for their court appearances, and 62.7 per cent went on to commit new crimes while out on bail – including the worst crimes of all: rape and murder. The algorithm had seen it all coming.

The researchers argued that, whichever way you use it, their algorithm vastly outperforms the human judge. And the numbers back them up. If you're wanting to incarcerate fewer people awaiting trial, the algorithm could help by consigning 41.8 per cent fewer defendants to jail while keeping the crime rate the same. Or, if you were happy with the current proportion of defendants given bail, then that's fine too: just by being more accurate at selecting which defendants to release, the algorithm could reduce the rate of skipping bail by 24.7 per cent.

These benefits aren't just theoretical. Rhode Island, where the courts have been using these kinds of algorithms for the last eight years, has achieved a 17 per cent reduction in prison populations and a 6 per cent drop in recidivism rates. That's hundreds of low-risk offenders who aren't unnecessarily stuck in prison, hundreds of crimes that haven't been committed. Plus, given that it costs over £30,000 a year to incarcerate one prisoner in the UK[26] – while in the United States spending a year in a high-security prison can cost about the same as going to Harvard[27] – that's hundreds of thousands of taxpayers' money saved. It's a win–win for everyone.

Or is it?

Finding Darth Vader

Of course, no algorithm can *perfectly* predict what a person is going to do in the future. Individual humans are too messy, irrational and impulsive for a forecast ever to be *certain* of what's going to happen next. They might give better predictions, but they will still make mistakes. The question is, what happens to all the people whose risk scores are wrong?

There are two kinds of mistake that the algorithm can make. Richard Berk, a professor of criminology and statistics at the University of Pennsylvania and a pioneer in the field of predicting recidivism, has a noteworthy way of describing them.

'There are good guys and bad guys,' he told me. 'Your algorithm is effectively asking: "Who are the Darth Vaders? And who are the Luke Skywalkers?"'

Letting a Darth Vader go free is one kind of error, known as a *false negative*. It happens whenever you fail to identify the risk that an individual poses.

Incarcerating Luke Skywalker, on the other hand, would be a *false positive*. This is when the algorithm incorrectly identifies someone as a high-risk individual.

These two kinds of error, false positive and false negative, are not unique to recidivism. They'll crop up repeatedly throughout this book. Any algorithm that aims to classify can be guilty of these mistakes.

Berk's algorithms claim to be able to predict whether someone will go on to commit a homicide with 75 per cent accuracy, which makes them some of the most accurate around.[28] When you consider how free we believe our will to be, that is a remarkably impressive level of accuracy. But even at 75 per cent, that's a lot of Luke Skywalkers who will be denied bail because they look like Darth Vaders from the outside.

The consequences of mislabelling a defendant become all the more serious when the algorithms are used in sentencing, rather than just decisions on bail or parole. This is a modern reality: recently, some US states have begun to allow judges to see a convicted offender's calculated risk score while deciding on their jail term. It's a development that has sparked a heated debate, and not without cause: it's one thing calculating whether to let someone out early, quite another to calculate how long they should be locked away in the first place.

Part of the problem is that deciding the length of a sentence involves consideration of a lot more than just the risk of a criminal re-offending – which is all the algorithm can help with. A judge also has to take into account the risk the offender poses to others, the deterrent effect the sentencing decision will have on other criminals,

the question of retribution for the victim and the chance of reha-
bilitation for the defendant. It's a lot to balance, so it's little wonder
that people raise objections to the algorithm being given too much
weight in the decision. Little wonder that people find stories like
that of Paul Zilly so deeply troubling.[29]

Zilly was convicted of stealing a lawnmower. He stood in front of
Judge Babler in Baron County, Wisconsin, in February 2013, knowing that his defence team had already agreed to a plea deal with the
prosecution. Both sides had agreed that, in his case, a long jail term
wasn't the best course of action. He arrived expecting the judge to
simply rubber-stamp the agreement.

Unfortunately for Zilly, Wisconsin judges were using a proprietary risk-assessment algorithm called COMPAS. As with the Idaho
budget tool in the 'Power' chapter, the inner workings of COM-
PAS are considered a trade secret. Unlike the budget tool, however,
the COMPAS code still isn't available to the public. What we do
know is that the calculations are based on the answers a defendant
gives to a questionnaire. This includes questions such as: 'A hungry
person has a right to steal, agree or disagree?' and: 'If you lived
with both your parents and they separated, how old were you at the
time?'[30] The algorithm was designed with the sole aim of predicting how likely a defendant would be to re-offend within two years,
and in this task had achieved an accuracy rate of around 70 per
cent.[31] That is, it would be wrong for roughly one in every three
defendants. None the less, it was being used by judges during their
sentencing decisions.

Zilly's score wasn't good. The algorithm had rated him as a high
risk for future violent crime and a medium risk for general recidivism. 'When I look at the risk assessment,' Judge Babler said in
court, 'it is about as bad as it could be.'

After seeing Zilly's score, the judge put more faith in the algorithm
than in the agreement reached by the defence and the prosecution,

rejected the plea bargain and doubled Zilly's sentence from one year in county jail to two years in a state prison.

It's impossible to know for sure whether Zilly deserved his high-risk score, although a 70 per cent accuracy rate seems a remarkably low threshold to justify using the algorithm to over-rule other factors.

Zilly's case was widely publicized, but it's not the only example. In 2003, Christopher Drew Brooks, a 19-year-old man, had consensual sex with a 14-year-old girl and was convicted of statutory rape by a court in Virginia. Initially, the sentencing guidelines suggested a jail term of 7 to 16 months. But, after the recommendation was adjusted to include his risk score (not built by COMPAS in this case), the upper limit was increased to 24 months. Taking this into account, the judge sentenced him to 18 months in prison.[32]

Here's the problem. This particular algorithm used age as a factor in calculating its recidivism score. Being convicted of a sex offence at such a young age counted against Brooks, even though it meant he was closer in age to the victim. In fact, had Brooks been 36 years old (and hence 22 years older than the girl) the algorithm would have recommended that he not be sent to prison at all.[33]

These are not the first examples of people trusting the output of a computer over their own judgement, and they won't be the last. The question is, what can you do about it? The Supreme Court of Wisconsin has its own suggestion. Speaking specifically about the danger of judges relying too heavily on the COMPAS algorithm, it stated: 'We expect that circuit courts will exercise discretion when assessing a COMPAS risk score with respect to each individual defendant.'[34] But Richard Berk suggests that might be optimistic: 'The courts are concerned about not making mistakes – especially the judges who are appointed by the public. The algorithm provides them a way to do less work while not being accountable.'[35]

There's another issue here. If an algorithm classifies someone as high risk and the judge denies them their freedom as a result, there is no way of knowing for sure whether the algorithm was seeing their future accurately. Take Zilly. Maybe he would have gone on to be violent. Maybe he wouldn't have. Maybe, being labelled as a high-risk convict and being sent to a state prison set him on a different path from the one he was on with the agreed plea deal. With no way to verify the algorithm's predictions, we have no way of knowing whether the judge was right to believe the risk score, no way of verifying whether Zilly was in fact a Vader or a Skywalker.

This is a problem without an easy solution. How do you persuade people to apply a healthy dose of common sense when it comes to using these algorithms? But even if you could, there's another problem with predicting recidivism. Arguably the most contentious of all.

Machine bias

In 2016, the independent online newsroom ProPublica, which first reported Zilly's story, looked in detail at the COMPAS algorithm and reverse-engineered the predictions it had made on the futures of over seven thousand real offenders in Florida,[36] in cases dating from 2013 or 2014. The researchers wanted to check the accuracy of COMPAS scores by seeing who had, in fact, gone on to re-offend. But they also wanted to see if there was any difference between the predicted risks for black and white defendants.

While the algorithm doesn't explicitly include race as a factor, the journalists found that not everyone was being treated equally within the calculations. Although the chances of the algorithm making an error were roughly the same for black or white offenders overall, it was making different kinds of mistakes for each racial group.

If you were one of the defendants who didn't get into trouble again after their initial arrest, a Luke Skywalker, the algorithm was twice as likely to mistakenly label you as high risk if you were black as it was if you were white. The algorithm's false positives were disproportionately black. Conversely, of all the defendants who did go on to commit another crime within the two years, the Darth Vaders, the white convicts were twice as likely to be mistakenly predicted as low risk by the algorithm as their black counterparts. The algorithm's false negatives were disproportionately white.

Unsurprisingly, the ProPublica analysis sparked outrage across the US and further afield. Hundreds of articles were written expressing sharp disapproval of the use of faceless calculations in human justice, rebuking the use of imperfect, biased algorithms in decisions that can have such a dramatic impact on someone's future. Many of the criticisms are difficult to disagree with – everyone deserves fair and equal treatment, regardless of who assesses their case, and the ProPublica study doesn't make things look good for the algorithm.

But let's be wary of our tendency to throw 'imperfect algorithms' away for a moment. Before we dismiss the use of algorithms in the justice system altogether, it's worth asking: What would you expect an unbiased algorithm to look like?

You'd certainly want it to make equally accurate predictions for black and white defendants. It also seems sensible to demand that what counts as 'high risk' should be the same for everyone. The algorithm should be equally good at picking out the defendants who are likely to re-offend, whatever racial (or other) group they belong to. Plus, as ProPublica pointed out, the algorithm should make the same kind of mistakes at the same rate for everyone – regardless of race.

None of these four statements seems like a particularly grand ambition. But there is, nevertheless, a problem. Unfortunately, some kinds of fairness are mathematically incompatible with others.

Let me explain. Imagine stopping people in the street and using an algorithm to predict whether each person will go on to commit a homicide. Now, since the vast majority of murders are committed by men (in fact, worldwide, 96 per cent of murderers are male)[37], if the murderer-finding algorithm is to make accurate predictions, it will necessarily identify more men than women as high risk.

Let's assume our murderer-detection algorithm has a prediction rate of 75 per cent. That is to say, three-quarters of the people the algorithm labels as high risk are indeed Darth Vaders.

Eventually, after stopping enough strangers, you'll have 100 people flagged by the algorithm as potential murderers. To match the perpetrator statistics, 96 of those 100 will necessarily be male. Four will be female. There's a picture below to illustrate. The men are represented by dark circles, the women shown as light grey circles.

Now, since the algorithm predicts correctly for both men and women at the same rate of 75 per cent, one-quarter of the females, and one-quarter of the males, will really be Luke Skywalkers: people who are incorrectly identified as high risk, when they don't actually pose a danger.

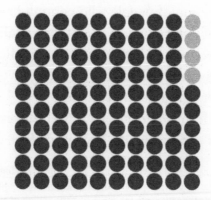

False positives

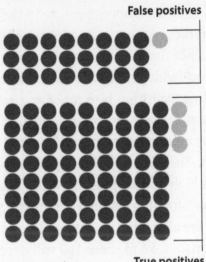

True positives

Once you run the numbers, as you can see from the second image here, more innocent men than innocent women will be incorrectly accused, just by virtue of the fact that men commit more murder than women.

This has nothing to do with the crime itself, or with the algorithm: it's just a mathematical certainty. The outcome is biased because reality is biased. More men commit homicides, so more men will be falsely accused of having the potential to murder.*

Unless the fraction of people who commit crimes is the same in every group of defendants, it is mathematically impossible to create a test which is equally accurate at prediction across the board *and* makes false positive and false negative mistakes at the same rate for every group of defendants.

Of course, African Americans have been subject to centuries of terrible prejudice and inequality. Because of this, they continue to

* An outcome like this can happen even if you're not explicitly using gender as a factor within the algorithm. As long as the prediction is based on factors that correlate with one group more than another (like a defendant's history of violent crime), this kind of unfairness can arise.

appear disproportionately at lower levels on the socio-economic scale and at higher levels in the crime statistics. There is also some evidence that – at least for some crimes in the US – the black population is disproportionately targeted by police. Marijuana use, for instance, happens in blacks and whites at the same rate, and yet arrest rates in African Americans can be up to eight times higher.[38] Whatever the reason for the disparity, the sad result is that rates of arrest are not the same across racial groups in the United States. Blacks are re-arrested more often than whites. The algorithm is judging them not on the colour of their skin, but on the all-too-predictable consequences of America's historically deeply unbalanced society. Until all groups are arrested at the same rate, this kind of bias is a mathematical certainty.

That is not to completely dismiss the ProPublica piece. Its analysis highlights how easily algorithms can perpetuate the inequalities of the past. Nor is it to excuse the COMPAS algorithm. Any company that profits from analysing people's data has a moral responsibility (if not yet a legal one) to come clean about its flaws and pitfalls. Instead, Equivant (formerly Northpointe), the company that makes COMPAS, continues to keep the insides of its algorithm a closely guarded secret, to protect the firm's intellectual property.[39]

There are options here. There's nothing inherent in these algorithms that means they have to repeat the biases of the past. It all comes down to the data you give them. We can choose to be 'crass empiricists' (as Richard Berk puts it) and follow the numbers that are already there, or we can decide that the status quo is unfair and tweak the numbers accordingly.

To give you an analogy, try doing a Google Images search for 'maths professor'. Perhaps unsurprisingly, you'll find the vast majority of images show white middle-aged men standing in front of chalk-covered blackboards. My search returned only one picture of a female in the top twenty images, which is a depressingly accurate reflection of reality: around 94 per cent of maths professors are male.[40] But however accurate the results might be, you could argue that using algorithms as a mirror to reflect the real world isn't always helpful, especially when the mirror is reflecting a present reality that only exists because of centuries of bias. Now, if it so chose, Google could subtly tweak its algorithm to prioritize images of female or non-white professors over others, to even out the balance a little and reflect the society we're aiming for, rather than the one we live in.

It's the same in the justice system. Effectively, using an algorithm lets us ask: what percentage of a particular group do we expect to be high risk in a perfectly fair society? The algorithm gives us the option to jump straight to that figure. Or, if we decide that removing all the bias from the judicial system at once isn't appropriate, we could instead ask the algorithm to move incrementally towards that goal over time.

There are also options in how you treat defendants with a high-risk score. In bail, where the risk of a defendant failing to appear at a future court date is a key component of the algorithm's prediction, the standard approach is to deny bail to anyone with a high-risk score. But the algorithm could also present an opportunity to find out why someone might miss a court date. Do they have access to suitable transport to get there? Are there issues with childcare that might prevent them attending? Are there societal imbalances that the algorithm could be programmed to alleviate rather than exacerbate?

The answers to these questions should come from the forums of open public debate and the halls of government rather than the boardrooms of private companies. Thankfully, the calls are getting louder for an algorithmic regulating body to control the industry. Just as the US Food and Drug Administration does for pharmaceuticals, it would test accuracy, consistency and bias behind closed doors and have the authority to approve or deny the use of a product on real people. Until then, though, it's incredibly important that organizations like ProPublica continue to hold algorithms to account. Just as long as accusations of bias don't end in calls for these algorithms to be banned altogether. At least, not without thinking carefully about what we'd be left with if they were.

Difficult decisions

This is a vitally important point to pick up on. If we threw the algorithms away, what kind of a justice system would remain? Because inconsistency isn't the only flaw from which judges have been shown to suffer.

Legally, race, gender and class should not influence a judge's decision. (Justice is supposed to be blind, after all.) And yet, while the vast majority of judges want to be as unbiased as possible, the evidence has repeatedly shown that they do indeed discriminate. Studies within the US have shown that black defendants, on average, will go to prison for longer,[41] are less likely to be awarded bail,[42] are more likely to be given the death penalty,[43] and once on death row are more likely to be executed.[44] Other studies have shown that men are treated more severely than women for the same crime,[45] and that defendants with low levels of income and education are given substantially longer sentences.[46]

Just as with the algorithm, it's not necessarily explicit prejudices that are causing these biased outcomes, so much as history repeating itself. Societal and cultural biases can simply arise as an automatic consequence of the way humans make decisions.

To explain why, we first need to understand a few simple things about human intuition, so let's leave the courtroom for a moment while you consider this question:

A bat and a ball cost £1.10 in total.
The bat costs £1 more than the ball.
How much does the ball cost?

This puzzle, which was posed by the Nobel Prize winning economist and psychologist Daniel Kahneman in his bestselling book *Thinking, Fast and Slow*,[47] illustrates an important trap we all fall into when thinking.

This question has a correct answer that is easy to see on reflection, but also an incorrect answer that immediately springs to mind. The answer you first thought of, of course, was 10p.*

Don't feel bad if you didn't get the correct answer (5p). Neither did 71.8 per cent of judges when asked.[48] Even those who do eventually settle on the correct answer have to resist the urge to go with their first intuition.

It's the competition between intuition and considered thought that is key to our story of judges' decisions. Psychologists generally agree that we have two ways of thinking. System 1 is automatic, instinctive, but prone to mistakes. (This is the system responsible for the answer of 10p jumping to mind in the puzzle above.) System 2 is slow, analytic, considered, but often quite lazy.[49]

If you ask a person how they came to a decision, it is System 2 that will articulate an answer – but, in the words of Daniel Kahneman, 'it often endorses or rationalizes ideas and feelings that were generated by System 1'.[50]

In this, judges are no different from the rest of us. They are human, after all, and prone to the same whims and weaknesses as we all are. The fact is, our minds just aren't built for robust, rational assessment of big, complicated problems. We can't easily weigh up the various factors of a case and combine everything together in a

* A ball at 10p would mean the bat was £1.10, making £1.20 in total.

logical manner while blocking the intuitive System 1 from kicking in and taking a few cognitive short cuts.

When it comes to bail, for instance, you might hope that judges were able to look at the whole case together, carefully balancing all the pros and cons before coming to a decision. But unfortunately, the evidence says otherwise. Instead, psychologists have shown that judges are doing nothing more strategic than going through an ordered checklist of warning flags in their heads. If any of those flags – past convictions, community ties, prosecution's request – are raised by the defendant's story, the judge will stop and deny bail.[51]

The problem is that so many of those flags are correlated with race, gender and education level. Judges can't help relying on intuition more than they should; and in doing so, they are unwittingly perpetuating biases in the system.

And that's not all. Sadly, we've barely scratched the surface of how terrible people are at being fair and unbiased judges.

If you've ever convinced yourself that an extremely expensive item of clothing was good value just because it was 50 per cent off (as I regularly do), then you'll know all about the so-called *anchoring effect*. We find it difficult to put numerical values on things, and are much more comfortable making comparisons between values than just coming up with a single value out of the blue. Marketers have been using the anchoring effect for years, not only to influence how highly we value certain items, but also to control the quantities of items we buy. Like those signs in supermarkets that say 'Limit of 12 cans of soup per customer'. They aren't designed to ward off soup fiends from buying up all the stock, as you might think. They exist to subtly manipulate your perception of how many cans of soup you need. The brain anchors with the number 12 and adjusts downwards. One study back in the 1990s showed that precisely such a sign could increase the average sale per customer from 3.3 tins of soup to 7.[52]

By now, you won't be surprised to learn that judges are also susceptible to the anchoring effect. They're more likely to award higher damages if the prosecution demands a high amount,[53] and hand down a longer sentence if the prosecutor requests a harsher punishment.[54] One study even showed that you could significantly influence the length of a sentence in a hypothetical case by having a journalist call the judge during a recess and subtly drop a suggested sentence into the conversation. ('Do you think the sentence in this case should be higher or lower than three years?')[55] Perhaps worst of all, it looks like you can tweak a judge's decision just by having them throw a dice before reviewing a case.[56] Even the most experienced judges were susceptible to this kind of manipulation.[57]

And there's another flaw in the way humans make comparisons between numbers that has an impact on the fairness of sentencing. You may have noticed this particular quirk of the mind yourself. The effect of increasing the volume on your stereo by one level diminishes the louder it plays; a price hike from £1.20 to £2.20 feels enormous, but an increase from £67 to £68 doesn't seem to matter; and time seems to speed up as you get older. It happens because humans' senses work in relative terms rather than in absolute values. We don't perceive each year as a fixed period of time; we experience each new year as a smaller and smaller fraction of the life we've lived. The size of the chunks of time or money or volume we perceive follows a very simple mathematical expression known as Weber's Law.

Put simply, Weber's Law states that the smallest change in a stimulus that can be perceived, the so-called 'Just Noticeable Difference', is proportional to the initial stimulus. Unsurprisingly, this discovery has also been exploited by marketers. They know exactly how much they can get away with shrinking a chocolate bar before customers notice, or precisely how much they can nudge up the price of an item before you'll think it's worth shopping around.

The problem in the context of justice is that Weber's Law influences the sentence lengths that judges choose. Gaps between sentences get bigger as the penalties get more severe. If a crime is marginally worse than something deserving a 20-year sentence, an additional 3 months, say, doesn't seem enough: it doesn't feel there's enough of a difference between a stretch of 20 years and one of 20 years and 3 months. But of course there is: 3 months in prison is still 3 months in prison, regardless of what came before. And yet, instead of adding a few months on, judges will jump to the next noticeably different sentence length, which in this case is 25 years.[58]

We know this is happening because we can compare the sentence lengths actually handed out to those Weber's Law would predict. One study from 2017 looked at over a hundred thousand sentences in both Britain and Australia and found that up to 99 per cent of defendants deemed guilty were given a sentence that fits the formula.[59]

'It doesn't matter what type of offence you'd committed,' Mandeep Dhami, the lead author of the study, told me, 'or what type of defendant you were, or which country you were being sentenced in, or whether you had been given a custodial or community sentence.' All that mattered was the number that popped into the judge's mind and felt about right.

Sadly, when it comes to biased judges, I could go on. Judges with daughters are more likely to make decisions more favourable to women.[60] Judges are less likely to award bail if the local sports team has lost recently. And one famous study even suggested that the time of day affects your chances of a favourable outcome.[61] Although the research has yet to be replicated,[62] and there is some debate over the size of the effect, there may be some evidence that taking the stand just before lunch puts you at a disadvantage: judges in the original study were most likely to award bail if they had just come back from a recess, and least likely when they were approaching a food break.

Another study showed that an individual judge will avoid making too many similar judgments in a row. Your chances, therefore, of being awarded bail drop off a cliff if four successful cases were heard immediately before yours.[63]

Some researchers claim, too, that our perceptions of strangers change depending on the temperature of a drink we're holding. If you're handed a warm drink just before meeting a new person, they suggest, you're more likely to see them as having a warmer, more generous, more caring personality.[64]

This long list is just the stuff we can measure. There are undoubtedly countless other factors subtly influencing our behaviour which don't lend themselves to testing in a courtroom.

Summing up

I'll level with you. When I first heard about algorithms being used in courtrooms I was against the idea. An algorithm will make mistakes, and when a mistake can mean someone loses their right to freedom, I didn't think it was responsible to put that power in the hands of a machine.

I'm not alone in this. Many (perhaps most) people who find themselves on the wrong side of the criminal justice system feel the same. Mandeep Dhami told me how the offenders she'd worked with felt about how decisions on their future were made. 'Even knowing that the human judge might make more errors, the offenders still prefer a human to an algorithm. They want that human touch.'

So, for that matter, do the lawyers. One London-based defence lawyer I spoke to told me that his role in the courtroom was to exploit the uncertainty in the system, something that the algorithm would make more difficult. 'The more predictable the decisions get, the less room there is for the art of advocacy.'

However, when I asked Mandeep Dhami how she would feel herself if she were the one facing jail, her answer was quite the opposite:

'I don't want someone to use intuition when they're making a decision about my future. I want someone to use a reasoned strategy. We want to keep judicial discretion, as though it is something so holy. As though it's so good. Even though research shows that it's not. It's not great at all.'

Like the rest of us, I think that judges' decisions should be as unbiased as possible. They should be guided by facts about the individual, not the group they happen to belong to. In that respect, the algorithm doesn't measure up well. But it's not enough to simply point at what's wrong with the algorithm. The choice isn't between a flawed algorithm and some imaginary perfect system. The only fair comparison to make is between the algorithm and what we'd be left with in its absence.

The more I've read, the more people I've spoken to, the more I've come to believe that we're expecting a bit too much of our human judges. Injustice is built into our human systems. For every Christopher Drew Brooks, treated unfairly by an algorithm, there are countless cases like that of Nicholas Robinson, where a judge errs on their own. Having an algorithm – even an imperfect algorithm – working with judges to support their often faulty cognition is, I think, a step in the right direction. At least a well-designed and properly regulated algorithm can help get rid of systematic bias and random error. You can't change a whole cohort of judges, especially if they're not able to tell you how they make their decisions in the first place.

Designing an algorithm for use in the criminal justice system demands that we sit down and think hard about exactly what the justice system is for. Rather than just closing our eyes and hoping for the best, algorithms require a clear, unambiguous idea of exactly what we want them to achieve and a solid understanding of the human failings they're replacing. It forces a difficult debate about precisely how a decision in a courtroom should be made. That's not going to be simple, but it's the key to establishing whether the algorithm can ever be good enough.

There are tensions within the justice system that muddy the waters and make these kinds of questions particularly difficult to answer. But there are other areas, slowly being penetrated by algorithms, where decisions are far less fraught with conflict, and the algorithm's objectives and positive contribution to society are far more clear-cut.

Medicine

IN 2015, A GROUP OF pioneering scientists conducted an unusual study on the accuracy of cancer diagnoses.[1] They gave 16 testers a touch-screen monitor and tasked them with sorting through images of breast tissue. The pathology samples had been taken from real women, from whom breast tissue had been removed by a biopsy, sliced thinly and stained with chemicals to make the blood vessels and milk ducts stand out in reds, purples and blues. All the tester had to do was decide whether the patterns in the image hinted at cancer lurking among the cells.

After a short period of training, the testers were set to work, with impressive results. Working independently, they correctly assessed 85 per cent of samples.

But then the researchers realized something remarkable. If they started pooling answers – combining votes from the individual testers to give an overall assessment on an image – the accuracy rate shot up to 99 per cent.

What was truly extraordinary about this study was not the skill of the testers. It was their identity. These plucky lifesavers were not oncologists. They were not pathologists. They were not nurses. They were not even medical students. They were pigeons.

Pathologists' jobs are safe for a while yet – I don't think even the scientists who designed the study were suggesting that doctors should be replaced by plain old pigeons. But the experiment did demonstrate an important point: spotting patterns hiding among clusters of cells is not a uniquely human skill. So, if a pigeon can manage it, why not an algorithm?

Pattern hunters

The entire history and practice of modern medicine is built on the finding of patterns in data. Ever since Hippocrates founded his school of medicine in ancient Greece some 2,500 years ago, observation, experimentation and the analysis of data have been fundamental to the fight to keep us healthy.

Before then, medicine had been – for the most part – barely distinguishable from magic. People believed that you fell ill if you'd displeased some god or other, and that disease was the result of an evil spirit possessing your body. As a result, the work of a physician would involve a lot of chanting and singing and superstition, which sounds like a lot of fun, but probably not for the person who was relying on it all to stop them from dying.

It wasn't as if Hippocrates single-handedly cured the world of irrationality and superstition for ever (after all, one rumour said he had a hundred-foot dragon for a daughter),[2] but he did have a truly revolutionary approach to medicine. He believed that the causes of disease were to be understood through rational investigation, not magic. By placing his emphasis on case reporting and observation, he established medicine as a science, justly earning himself a reputation as 'the father of modern medicine'.[3]

While the scientific explanations that Hippocrates and his colleagues came up with don't exactly stand up to modern scrutiny (they believed that health was a harmonious balance of blood, phlegm, yellow bile and black bile),[4] the conclusions they drew from their data certainly do.[5] (They were the first to give us insights such as: 'Patients who are naturally fat are apt to die earlier than those who are slender.') It's a theme that is found throughout the ages. Our scientific understanding may have taken many wrong turns along the way, but progress is made through our ability to find patterns, classify symptoms and use these observations to predict what the future holds for a patient.

Medical history is packed with examples. Take fifteenth-century China, when healers first realized they could inoculate people against smallpox. After centuries of experimentation, they found a pattern that they could exploit to reduce the risk of death from this illness by a factor of ten. All they had to do was find an individual with a mild case of the disease, harvest their scabs, dry them, crush them and blow them into the nose of a healthy person.[6] Or the medical golden age of the nineteenth century, as medicine adopted increasingly scientific methods, and looking for patterns in data became integral to the role of a physician. One of these physicians was the Hungarian Ignaz Semmelweis, who in the 1840s noticed something startling in the data on deaths on maternity wards. Women who gave birth in wards staffed by doctors were five times more likely to fall ill to sepsis than those in wards run by midwives. The data also pointed towards the reason why: doctors were dissecting dead bodies and then immediately attending to pregnant women without stopping to wash their hands.[7]

What was true of fifteenth-century China and nineteenth-century Europe is true today of doctors all over the world. Not just when studying diseases in the population, but in the day-to-day role of a primary care giver, too. Is this bone broken or not? Is this head-ache perfectly normal or a sign of something more sinister? Is it worth prescribing a course of antibiotics to make this boil go away? All are questions of pattern recognition, classification and prediction. Skills that algorithms happen to be very, very good at.

Of course, there are many aspects of being a doctor that an algorithm will probably never be able to replicate. Empathy, for one thing. Or the ability to support patients through social, psychological, even financial difficulties. But there are some areas of medicine where algorithms can offer a helping hand. Especially in the roles where medical pattern recognition is found in its purest form and classification and prediction are prized almost to the exclusion of all else. Especially in an area like pathology.

Pathologists are the doctors a patient rarely meets. Whenever you have a blood or tissue sample sent off to be tested, they're the ones who, sitting in some distant laboratory, will examine your sample and write the report. Their role sits right at the end of the diagnostic line, where skill, accuracy and reliability are crucially important. They are often the ones who say whether you have cancer or not. So, if the biopsy they're analysing is the only thing between you and chemotherapy, surgery or worse, you want to be sure they're getting it right.

And their job isn't easy. As part of their role, the average pathologist will examine hundreds of slides a day, each containing tens of thousands – sometimes hundreds of thousands – of cells suspended between the small glass plates. It's the hardest game of *Where's Wally?* imaginable. Their job is to meticulously scan each sample, looking for tiny anomalies that could be hiding anywhere in the vast galaxy of cells they see beneath the microscope's lens.

'It's an impossibly hard task,' says Andy Beck,[8] a Harvard pathologist and founder of PathAI, a company created in 2016 that creates algorithms to classify biopsy slides. 'If each pathologist were to look very carefully at five slides a day, you could imagine they might achieve perfection. But that's not the real world.'

It certainly isn't. And in the real world, their job is made all the harder by the frustrating complexities of biology. Let's return to the example of breast cancer that the pigeons were so good at spotting. Deciding if someone has the disease isn't a straight yes or no. Breast cancer diagnoses are spread over a spectrum. At one end are the benign samples where normal cells appear exactly as they should be. At the other end are the nastiest kind of tumours – invasive carcinomas, where the cancer cells have left the milk ducts and begun to grow into the surrounding tissue. Cases that are at these extremes are relatively easy to spot. One recent study showed that pathologists manage to correctly diagnose 96 per cent of straightforward malignant specimens, an accuracy roughly equivalent to what the flock of pigeons managed when given a similar task.[9]

But in between these extremes – between totally normal and obviously horribly malignant – there are several other, more ambiguous categories, as shown in the diagram below. Your sample could have a group of atypical cells that look a bit suspicious, but aren't necessarily anything to worry about. You could have pre-cancerous growths that may or may not turn out to be serious. Or you could have cancer that has not yet spread outside the milk ducts (so-called ductal carcinoma in situ).

Which particular category your sample happens to be judged as falling into will probably have an enormous impact on your treatment. Depending on where your sample sits on the line, your doctor could suggest anything from a mastectomy to no intervention at all.

The problem is, distinguishing between these ambiguous categories can be extremely tricky. Even expert pathologists can disagree

| Normal duct | Intraductal hyperplasia | Intraductal hyperplasia with atypia | Intraductal carcinoma in situ | Invasive ductal cancer |

on the correct diagnosis of a single sample. To test how much the doctors' opinions varied, one 2015 study took 72 biopsies of breast tissue, all of which were deemed to contain cells with benign abnormalities (a category towards the middle of the spectrum) and asked 115 pathologists for their opinion. Worryingly, the pathologists only came to the same diagnosis 48 per cent of the time.[10]

Once you're down to 50–50, you might as well be flipping a coin for your diagnosis. Heads and you could end up having an unnecessary mastectomy (costing you hundreds of thousands of dollars if you live in the United States). Tails and you could miss a chance to address your cancer at its earliest stage. Either way, the impact can be devastating.

When the stakes are this high, accuracy is what matters most. So can an algorithm do better?

Machines that see

Until recently, creating an algorithm that could recognize anything at all in an image, let alone cancerous cells, was considered a notoriously tricky challenge. It didn't matter that understanding pictures comes so easily to humans; explaining precisely *how* we were doing it proved an unimaginably difficult task.

To understand why, imagine writing instructions to tell a computer whether or not a photo has a dog in it. You could start off with the obvious stuff: if it has four legs, if it has floppy ears, if it has fur and so on. But what about those photos where the dog is sitting down? Or the ones in which you can't see all the legs? What about dogs with pointy ears? Or cocked ears? Or the ones not facing the camera? And how does 'fur' look different from a fluffy carpet? Or the wool of a sheep? Or grass?

Sure, you could work all of these in as extra instructions, running through every single possible type of dog ear, or dog fur, or sitting position, but your algorithm will soon become so enormous it'll be entirely unworkable, before you've even begun to distinguish dogs from other four-legged furry creatures. You need to find another way. The trick is to shift away from the rule-based paradigm and use something called a 'neural network'.[11]

You can imagine a neural network as an enormous mathematical structure that features a great many knobs and dials. You feed your picture in at one end, it flows through the structure, and out at the other end comes a guess as to what that image contains. A probability for each category: Dog; Not dog.

At the beginning, your neural network is a complete pile of junk. It starts with no knowledge – no idea of what is or isn't a dog. All the dials and knobs are set to random. As a result, the answers it provides are all over the place – it couldn't accurately recognize an image if its power source depended on it. But with every picture you feed into it, you tweak those knobs and dials. Slowly, you train it.

In goes your image of a dog. After every guess the network makes, a set of mathematical rules works to adjust all the knobs until the prediction gets closer to the right answer. Then you feed in another image, and another, tweaking every time it gets something wrong; reinforcing the paths through the array of knobs that lead to success and fading those that lead to failure. Information about what makes one dog picture similar to another dog picture propagates backwards through the network. This continues until – after hundreds and thousands of photos have been fed through – it gets as few wrong as possible. Eventually you can show it an image that it has never seen before and it will be able to tell you with a high degree of accuracy whether or not a dog is pictured.

The surprising thing about neural networks is that their operators usually don't understand how or why the algorithm reaches its conclusions. A neural network classifying dog pictures doesn't work by picking up on features that you or I might recognize as dog-like. It's not looking for a measure of 'chihuahua-ness' or 'Great Dane-ishness' – it's all a lot more abstract than that: picking up on patterns of edges and light and darkness in the photos that don't make a lot of sense to a human observer (have a look at the image recognition example in the 'Power' chapter to see what I mean). Since the process is difficult for a human to conceptualize, it means the operators only know that they've tuned up their algorithm to get the answers right; they don't necessarily know the precise details of how it gets there.

This is another 'machine-learning algorithm', like the random forests we met in the 'Justice' chapter. It goes beyond what the operators program it to do and learns itself from the images it's given. It's this ability to learn that endows the algorithm with 'artificial intelligence'. And the many layers of knobs and dials also give the network a deep structure, hence the term 'deep learning'.

Neural networks have been around since the middle of the twentieth century, but until quite recently we've lacked the widespread access to really powerful computers necessary to get the best out of them. The world was finally forced to sit up and take them seriously in 2012 when computer scientist Geoffrey Hinton and two of his students entered a new kind of neural network into an image recognition competition.[12] The challenge was to recognize – among other things – dogs. Their artificially intelligent algorithm blew the best of its competitors out of the water and kicked off a massive renaissance in deep learning.

An algorithm that works without our knowing how it makes its decisions might sound like witchcraft, but it might not be all that dissimilar from how we learn ourselves. Consider this comparison. One team recently trained an algorithm to distinguish between photos of wolves and pet huskies. They then showed how, thanks to the way it had tuned its own dials, the algorithm wasn't using anything to do with the dogs as clues at all. It was basing its answer on whether the picture had snow in the background. Snow: wolf. No snow: husky.[13]

Shortly after their paper was published, I was chatting with Frank Kelly, a professor of mathematics at Cambridge University, who told me about a conversation he'd had with his grandson. He was walking the four-year-old to nursery when they passed a husky. His grandson remarked that the dog 'looked like' a wolf. When Frank asked how he knew that it wasn't a wolf, he replied, 'Because it's on a lead.'

An AI alliance

There are two things you want from a good breast cancer screening algorithm. You want it to be *sensitive* enough to pick up on the abnormalities present in all the breasts that have tumours, without skipping over the pixels in the image and announcing them as clear. But you also want it to be *specific* enough not to flag perfectly normal breast tissue as suspicious.

We've met the principles of sensitivity and specificity before, in the 'Justice' chapter. They are close cousins of *false negatives* and *false positives* (or Darth Vader and Luke Skywalker – which, if you ask me, is how they should be officially referred to in the scientific literature). In the context we're talking about here, a false positive occurs whenever a healthy woman is told she has breast cancer, and a false negative when a woman with tumours is given the all-clear. A specific test will have hardly any false positives, while a sensitive one has few false negatives. It doesn't matter what context your algorithm is working in – predicting recidivism, diagnosing breast cancer or (as we'll see in the 'Crime' chapter) identifying patterns of criminal activity – the story is always the same. You want as few false positives and false negatives as possible.

The problem is that refining an algorithm often means making a choice between sensitivity and specificity. If you focus on improving one, it often means a loss in the other. If, for instance, you decided to prioritize the complete elimination of false negatives, your algorithm could flag every single breast it saw as suspicious. That would score 100 per cent sensitivity, which would certainly satisfy your objective. But it would also mean an awful lot of perfectly healthy people undergoing unnecessary treatment. Or say you decided to prioritize the complete elimination of false positives. Your algorithm would wave everyone through as healthy, thus earning a 100 per cent score on specificity. Wonderful! Unless you're one of the women with tumours that the algorithm just disregarded.

Interestingly, human pathologists don't tend to have problems with specificity. They almost never mistakenly identify cells as cancerous when they're not. But people do struggle a little with sensitivity. It's worryingly easy for us to miss tiny tumours – even obviously malignant ones.

These human weaknesses were highlighted in a recent challenge that was designed to pit human against algorithm. Computer teams from around the world went head-to-head with a pathologist to find all the tumours within four hundred slides, in a competition known as CAMELYON16. To make things easier, all the cases were at the two extremes: perfectly normal tissue or invasive breast cancer. There was no time constraint on the pathologist, either: they could take as long as they liked to wade through the biopsies. As expected, the pathologist generally got the overall diagnosis right (96 per cent accuracy)[14] – and without identifying a single false positive in the process. But they also missed a lot of the tiny cancer cells hiding among the tissue, only managing to spot 73 per cent of them in 30 hours of looking.

The sheer number of pixels that needed checking wasn't necessarily the problem. People can easily miss very obvious anomalies even when looking directly at them. In 2013, Harvard researchers

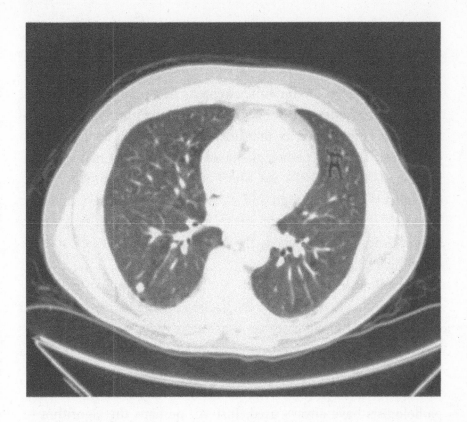

secretly hid an image of a gorilla in a series of chest scans and asked 24 unsuspecting radiologists to check the images for signs of cancer. Eighty-three per cent of them failed to notice the gorilla, despite eye-tracking showing that the majority were literally looking right at it.[15] Try it yourself with the picture above.[16]

Algorithms have the opposite problem. They will eagerly spot anomalous groups of cells, even perfectly healthy ones. During CAMELYON16, for instance, the best neural network entered managed to find an impressive 92.4 per cent of the tumours,[17] but in doing so it made eight false-positive mistakes per slide by incorrectly flagging normal groups of cells as suspicious. With such a low specificity, the current state-of-the art algorithms are definitely leaning towards the 'everyone has breast cancer!' approach to diagnosis and are just not good enough to create their own pathology reports yet.

The good news, though, is that we're not asking them to. Instead, the intention is to combine the strengths of human and machine. The algorithm does the donkey-work of searching the enormous amount of information in the slides, highlighting a few key areas of interest. Then the pathologist takes over. It doesn't matter if the machine is flagging cells that aren't cancerous; the human expert can quickly check through and eliminate anything that's normal. This kind of algorithmic pre-screening partnership not only saves a lot of time, it also bumps up the overall accuracy of diagnosis to a stunning 99.5 per cent.[18]

Marvellous as all this sounds, the fact is that human pathologists have always been good at diagnosing aggressive cancerous tumours. The difficult cases are those ambiguous ones in the middle, where the distinction between cancer and not cancer is more subtle. Can the algorithms help here too? The answer is (probably) yes. But not by trying to diagnose using the tricky categories that pathologists have always used. Instead, perhaps the algorithm – which is so much better at finding anomalies hidden in tiny fragments of data – could offer a better way to diagnose altogether. By doing something that human doctors can't.

The Nun Study

In 1986, an epidemiologist from the University of Kentucky named David Snowden managed to persuade 678 nuns to give him their brains. The nuns, who were all members of the School Sisters of Notre Dame, agreed to participate in Snowden's extraordinary scientific investigation into the causes of Alzheimer's disease.

Each of these women, who at the beginning of the study were aged between 75 and 103 years old, would take a series of memory tests every year for the rest of their lives. Then, when they died, their brains would be donated to the project. Regardless of whether they had symptoms of dementia or not, they promised to allow Snowden's team to remove their most precious organ and analyse it for signs of the disease.[19]

The nuns' generosity lead to the creation of a remarkable dataset. Since none of them had children, smoked, or drank very much, the scientists were able to rule out many of the external factors believed to raise the likelihood of contracting Alzheimer's. And since they all lived a similar lifestyle, with similar access to healthcare and social support, the nuns effectively provided their own experimental control.

All was going well when, a few years into the study, the team discovered this experimental group offered another treasure trove of data they could tap into. As young women, many of the now elderly nuns had been required to submit a handwritten autobiographical essay to the Sisterhood before they were allowed to take their vows. These essays were written when the women were – on average – only 22 years old, decades before any would display any symptoms of dementia. And yet, astonishingly, the scientists discovered clues in their writing that predicted what would happen to them far in the future.

The researchers analysed the language in each of the essays for its complexity and found a connection between how articulate the nuns were as young women and their chances of developing dementia in old age.

For example, here is a one-sentence extract from a nun who maintained excellent cognitive ability throughout her life:

> After I finished the eighth grade in 1921 I desired to become an aspirant at Mankato but I myself did not have the courage to ask the permission of my parents so Sister Agreda did it in my stead and they readily gave their consent.

Compare this to a sentence written by a nun whose memory scores steadily declined in her later years:

After I left school, I worked in the post-office.

The association was so strong that the researchers could predict which nuns might have dementia just by reading their letters. Ninety per cent of the nuns who went on to develop Alzheimer's had 'low linguistic ability' as young women, while only 13 per cent of the nuns who maintained cognitive ability into old age got a 'low idea density' score in their essays.[20]

One of the things this study highlights is the incredible amount we still have to learn about our bodies. Even knowing that this connection might exist doesn't tell us *why*. (Is it that good education staves off dementia? Or that people with the propensity to develop Alzheimer's feel more comfortable with simple language?) But it might suggest that Alzheimer's can take several decades to develop.

More importantly, for our purposes, it demonstrates that subtle signals about our future health can hide in the tiniest, most unexpected fragments of data – years before we ever show symptoms of an illness. It hints at just how powerful future medical algorithms that can dig into data might be. Perhaps, one day, they'll even be able to spot the signs of cancer years before doctors can.

Powers of prediction

In the late 1970s, a group of pathologists in Ribe County, Denmark, started performing double mastectomies on a group of corpses. The deceased women ranged in age from 22 to 89 years old, and 6 of them, out of 83, had died of invasive breast cancer. Sure enough, when the researchers prepared the removed breasts for examination by a pathologist – cutting each into four pieces and then slicing the tissue thinly on to slides – these 6 samples revealed the hallmarks of the disease. But to the researchers' astonishment, of the remaining 77 women – who had died from completely unrelated causes, including heart disease and car accidents – almost a quarter had the warning signs of breast cancer that pathologists look out for in living patients.

Without ever showing any signs of illness, 14 of the women had in situ cancer cells that had never developed beyond the milk ducts or glands. Cells that would be considered malignant breast cancer if the women were alive. Three had atypical cells that would also be flagged as a matter of concern in a biopsy, and one woman actually had invasive breast cancer without having any idea of it when she died.[21]

These numbers were surprising, but the study wasn't a fluke. Other researchers have found similar results. In fact, some estimate that, at any one time, around 9 per cent of women could be unwittingly walking around with tumours in their breasts[22] – about ten times the proportion who actually get diagnosed with breast cancer.[23]

So what's going on? Do we have a silent epidemic on our hands? According to Dr Jonathan Kanevsky, a medical innovator and resident in surgery at McGill University in Montreal, the answer is no. At least, not really. Because the presence of cancer isn't necessarily a problem:

> If somebody has a cancer cell in their body, the chances are their immune system will identify it as a mutated cell and just attack it and kill it – that cancer will never grow into something scary. But sometimes the immune system messes up, meaning the body supports the growth of the cancer, allowing it to develop. At that point cancer can kill.[24]

Not all tumours are created equally. Some will be dealt with by your body, some will sit there quite happily until you die, some could develop into full-blown aggressive cancer. The trouble is, we often have very little way of knowing which will turn out to be which.

And that's why these tricky categories between benign and horribly malignant can be such a problem. They are the only classifications that doctors have to work with, but if a doctor finds a group of cells in your biopsy that seem a bit suspicious, the label they choose can only help describe what lies within your tissue now. That isn't necessarily much help when it comes to giving you clues to your future. And it's the future, of course, that worried patients are most concerned about.

The result is that people are often overly cautious in the treatment they choose. Take in situ cancers, for instance. This category sits towards the more worrying end of the spectrum, where a cancerous growth is present, but hasn't yet spread to the surrounding tissue. Serious as this sounds, only around one in ten 'in situ' cancers will turn into something that could kill you. None the less, a quarter of women who receive this diagnosis in the United States will undergo a full mastectomy – a major operation that's physically and often emotionally life-changing.[25]

The fact is, the more aggressively you screen for breast cancer, the more you affect women who otherwise would've happily got on with their lives, oblivious to their harmless tumours. One independent UK panel concluded that for every 10,000 women who'll be invited to a mammogram screening in the next 20 years, 43 deaths from breast cancer will be prevented. And a study published in the *New England Journal of Medicine* concluded that for every 100,000 who attend routine mammogram screenings, tumours that could become life-threatening will be detected in 30 women.[26] But – depending which set of statistics you use – three or four times as many women will be over-diagnosed, receiving treatment for tumours that were never going to put their lives in danger.[27]

This problem of over-diagnosis and over-treatment is hard to solve when you're good at detecting abnormalities but not good at predicting how they'll develop. Still, there might be hope. Perhaps – just as with the essay-writing nuns – tiny clues to the way someone's health will turn out years in the future can be found hiding in their past and present data. If so, winkling out this information would be a perfect job for a neural network.

In an arena where doctors have struggled for decades to discover *why* one abnormality is more dangerous than another, an algorithm that isn't taught what to look for could come into its own. Just as long as you can put together a big enough set of biopsy slides (including some samples of tumours that eventually metastasized – spread to other parts of the body – as well as some that didn't) to train your neural network, it could hunt for hidden clues to your future health blissfully free of any of the prejudice that can come with theory. As Jonathan Kanevsky puts it: 'It would be up to the algorithm to determine the unique features within each image that correspond to whether the tumour becomes metastatic or not.'[28]

With an algorithm like that, the category your biopsy fitted into would become far less important. You wouldn't need to bother with why or how, you could just jump straight to the information that matters: do you need treatment or not?

The good news is, work on such an algorithm has already begun. Andy Beck, the Harvard pathologist and CEO of PathAI we met earlier, recently let his algorithm loose on a series of samples from patients in the Netherlands and found that the best predictors of patient survival were to be found not in the cancer itself but in other abnormalities in the adjacent tissue.[29] This is an important development – a concrete example of the algorithms themselves driving the research forwards, proving they can find patterns that improve our powers of prediction.

And, of course, there's now an incredible wealth of data we can draw on. Thanks to routine mammogram screening around the world, we probably have more images of breast tissue than of any other organ in the body. I'm not a pathologist, but every expert I've spoken to has convinced me that being able to confidently predict whether a troublesome sample will become cancerous is well within our sights. There's a very real chance that by the time this book is published in paperback, someone somewhere will have actually made this world-changing idea a reality.

Digital diagnosis

These ideas apply well beyond breast cancer. The neural networks that Andy Beck and others are building don't particularly care what they're looking at. You could ask them to categorize anything: dogs, hats, cheeses. As long as you're letting them know when they're getting it right and wrong, they'll learn. And now that this family of algorithms are good enough to be usable, they're having an impact on all sorts of areas of modern medicine.

One big recent success, for instance, comes from the Google Brain team, who have built an algorithm that screens for the world's biggest cause of preventable blindness – diabetic retinopathy. It's a disease that affects the blood vessels in the light-sensitive areas of the eye. If you know you have it, you can be given injections to save your sight, but if it's not caught early it can lead to irreversible blindness. In India, where access to experts capable of diagnosing the condition is limited, 45 per cent of people with diabetic retinopathy will lose some of their sight before they know they have the disease. The Google team's algorithm, which was built in a collaboration with doctors from India, is now just as good at diagnosing the condition as a human ophthalmologist.

Similarly, there are algorithms that look for cardiovascular diseases in the heart,[30] emphysema in the lungs,[31] strokes in the brain[32] and melanomas on the skin.[33] There are even systems that diagnose polyps during a live colonoscopy in real time.

The fact is, if you can take a picture of it and stick a label on it, you can create an algorithm to find it. And you're likely to end up with a more accurate (and possibly earlier) diagnosis than any human doctor could manage.

But what about the messier forms of medical data? Can the success of these algorithms be taken further, to something beyond these kinds of highly specialized, narrowly focused tasks? Could a machine search for meaning in your doctor's scribbled notes, for instance? Or pick up on tiny clues in how you describe the pain you're experiencing?

How about the ultimate healthcare science-fiction fantasy, where a machine in your doctor's surgery would carefully listen to your symptoms and analyse your medical history? Can we dare to imagine a machine that has a mastery of every scrap of cutting-edge medical research? One that offers an accurate diagnosis and perfectly tailored treatment plan?

In short, how about something a little like IBM's Watson?

Elementary, my dear

In 2004, Charles Lickel was tucking into a steak dinner with some colleagues in a New York restaurant. Partway through the meal, the dining area began to empty. Intrigued, Charles followed the crowd of diners and found them huddled around a television, eagerly watching the popular game show *Jeopardy*. The famous *Jeopardy* champion Ken Jennings had a chance to hold on to his record-breaking six-month winning streak and the diners didn't want to miss it.[34]

Charles Lickel was the vice-president of software at IBM. For the past few years, ever since Deep Blue had beaten Garry Kasparov at chess, the IBM bosses had been nagging Charles to find a new challenge worthy of the company's attention. As he stood in the New York restaurant, watching the diners captivated by this human *Jeopardy* champion, Lickel began to wonder if a machine could be designed to beat him.

It wouldn't be easy. The machine known as 'Watson' that Charles imagined in that restaurant took seven long years to build. But eventually Watson would challenge Ken Jennings on a special episode of *Jeopardy* and convincingly defeat him at the game he'd made his own. In the process, IBM would set themselves on the path to their attempt to build the world's first all-singing, all-dancing diagnosis machine. We'll come back to that in a moment. But first, let me walk you through some of the key ideas behind the *Jeopardy*-winning machine that formed the basis of the medical diagnosis algorithms.

For those who've never heard of it, *Jeopardy* is a well-known American game show in the form of a kind of reverse general knowledge quiz: the contestants are given clues in the form of answers and have to phrase their responses in the form of questions. For instance, in the category 'self-contradictory words' a clue might be:

A fastener to secure something; or it could be to bend, warp & give way suddenly with heat or pressure.

The algorithmic player would have to learn to work through a few layers to get to the correct response: 'What does "buckle" mean?' First, Watson would need to understand language well enough to derive meaning from the question and realize that 'fastener', 'secure', 'bend', 'warp' and 'give way suddenly' are all separate elements of the clue. This, in itself, is an enormous challenge for an algorithm.

But that was only the first step. Next, Watson would need to hunt for potential candidates that fitted each of the clues. 'Fastener' might conjure up all manner of potential answers: 'clasp', 'button', 'pin' and 'tie', as well as 'buckle', for instance. Watson needs to consider each possibility in turn and measure how it fits with the other clues. So while you're unlikely to find evidence of 'pin' being associated with the clues 'bend' and 'warp', the word 'buckle' certainly is, which increases Watson's confidence in it as a possible answer. Eventually, once all of the evidence has been combined, Watson has to put its imaginary money where its metaphorical mouth is by choosing a single response.

Now, the challenge of playing *Jeopardy* is rather more trivial than that of diagnosing disease, but it does require some of the same logical mechanisms. Imagine you go to the doctor complaining of unintentional weight loss and stomach aches, plus a bit of heartburn for good measure. In analogy with playing *Jeopardy*, the challenge is to find potential diagnoses (responses) that might explain the symptoms (clues), look for further evidence on each and update the confidence in a particular answer as more information becomes available. Doctors call this differential diagnosis. Mathematicians call it Bayesian inference.*

Even after Watson the quiz champion had been successfully created, building Watson the medical genius was no simple task. None the less, when IBM went public with their plans to move into healthcare, they didn't hold back from making grand promises. They told the world that Watson's ultimate mission was to 'eradicate cancer',[35] and hired the famous actor John Hamm to boast that it was 'one of the most powerful tools our species has created'.

It's certainly a vision of a medical utopia to inspire us all. Except – as you probably know already – Watson hasn't quite lived up to the hype.

First a prestigious contract with the University of Texas M. D. Anderson Cancer Center was terminated in 2016. Rumour had it that even after they had shelled out $62 million on the technology[36]

* More on Bayes in the 'Cars' chapter.

and spent four years working with it, Watson still wasn't able to do anything beyond heavily supervised pilot tests. Then, in late September 2017, an investigation by the health news website STAT reported that Watson was 'still struggling with the basic step of learning about different forms of cancer.'[37]

Ouch.

To be fair, it wasn't all bad news. In Japan, Watson did manage to diagnose a woman with a rare form of leukaemia, when doctors hadn't.[38] And analysis by Watson led to the discovery of five genes linked to motor neurone disease, or ALS.[39] But overall, IBM's programmers haven't quite managed to deliver on the promises of their excitable marketing department.

It's hard not to feel sympathy for anyone trying to build this kind of machine. It's theoretically possible to construct one that can diagnose disease (and even offer patients sensible treatment plans), and that's an admirable goal to have. But it's also really difficult. Much more difficult than playing *Jeopardy*, and much, much more difficult than recognizing cancerous cells in an image.

An all-purpose diagnostic machine might seem only a simple logical step on from those cancer-spotting image-based algorithms we met earlier, but those algorithms have a big advantage: they get to examine the actual cells that might be causing a problem. A diagnostic machine, by contrast, only gets information several steps removed from the underlying issue. Maybe a patient has pins and needles caused by a muscle spasm caused by a trapped nerve caused by too much heavy lifting. Or maybe they have blood in their stool caused by haemorrhoids caused by constipation caused by poor diet. An algorithm (or a doctor) has to take a single symptom and trace the route backwards to an accurate diagnosis. That is what Watson had to do. It is a monumentally difficult task.

And there are other problems too.

Remember that dog/wolf neural network? Training that was easy. All the programmers had to do was find a stack of photos labelled 'dog' or 'wolf' and feed them in. The dataset was simple and not ambiguous. 'But', as Thomas Fuchs, a computational pathologist, told *MIT Technology Review*, 'in a specialized domain in medicine, you might need experts trained for decades to properly label the information you feed to the computer.'[40]

That might be a surmountable problem for a really focused question (such as sorting breast cancer pathology slides into 'totally benign' and 'horribly malignant'). But an all-seeing diagnostic machine like Watson would need to understand virtually every possible disease. This would require an army of incredibly highly qualified human handlers prepared to feed it information about different patients and their specific characteristics for a very long time. And generally speaking, those people tend to have other stuff on their plate – like actually saving lives.

And then we come to the final problem. The most difficult of all to overcome.

The trouble with data

Tamara Mills was just a baby when her parents first noticed something wasn't right with her breathing. By the time she was nine months old, doctors had diagnosed her with asthma – a condition that affects 5.4 million people in the UK and 25 million in the US.[41] Although Tamara was younger than most sufferers, her symptoms were perfectly manageable in the early years of her life and she grew up much like any other kid with the condition, spending her childhood playing by the sea in the north of England (although always with an inhaler to hand).

When she was 8, Tamara caught a nasty bout of swine flu. It would prove to be a turning point for her health. From that point on, one chest infection followed another, and another. Sometimes, during her asthma attacks, her lips would turn blue. But no matter how often Tamara and her mother went to her doctor and the local hospital, no matter how often her parents complained that she was getting though her supply of inhalers faster than they could be prescribed,[42] none of the doctors referred her to a specialist.

Her family and teachers, on the other hand, realized things were getting serious. After two near-fatal attacks that put Tamara in hospital, she was excused from PE lessons at school. When the stairs at home became too much to manage, she went to live with her grandparents in their bungalow.

On 10 April 2014, Tamara succumbed to yet another chest infection. That night her grandfather found her struggling for breath. He called an ambulance and tried his best to help her with two inhalers and an oxygen tank. Her condition continued to deteriorate. Tamara died later that night, aged just 13 years old.

Asthma is not usually a fatal condition, and yet 1,200 people across the UK die from it every year, 26 of whom are children.[43] It's estimated that two-thirds of those deaths are preventable – as was Tamara's. But that prevention depends entirely on the warning signs being spotted and acted upon.

In the four years leading up to her final and fatal attack, Tamara visited her local doctors and hospital no fewer than 47 times. Her treatment plan clearly wasn't working, and yet each time she saw a healthcare professional, they only treated the immediate problem. No one looked at the bigger picture. No one spotted the pattern emerging in her visits; no one noticed that her condition was steadily deteriorating; no one suggested it was time to try something new.[44]

There was a reason for this. Believe it or not (and anyone who lives here probably will), the UK's National Health Service doesn't link its healthcare records together as a matter of standard practice.

If you find yourself in an NHS hospital, the doctors won't know anything about any visits you've made to your local GP. Many records are still kept on paper, which means the method of sharing them between doctors hasn't changed for decades. It's one of the reasons why the NHS holds the dubious title of the world's biggest purchaser of fax machines.[45]

Crazy as this sounds, we're not alone. The United States has a plethora of private doctors and large hospital networks that are not connected to each other; and while other countries, such as Germany, have begun to build electronic patient records, they are a long way from being the norm around the world. For Tamara Mills, the lack of a single, connected medical history meant that it was impossible for any individual doctor to fully understand the severity of her condition. Any solution to this profound shortcoming will come sadly too late for Tamara, but it remains a huge challenge for the future of healthcare. A machine like Watson could help save any number of Tamaras, but it'll only be able to find patterns in the data if that data is collected, collated and connected.

There is a stark contrast between the rich and detailed datasets owned by data brokers and the sparse and disconnected datasets found in healthcare. For now, medical data is a mess. Even when our detailed medical histories are stored in a single place (which they often aren't), the data itself can take so many forms that it's virtually impossible to connect the information in a way that's useful to an algorithm. There are scans to consider, reports to include, charts, prescriptions, notes: the list goes on. Then you have the problem of how the written data is recorded. You need to be able to understand all the acronyms and abbreviations, decipher the handwriting, identify the possibilities for human error. And that's before you even get to symptoms. Does this person mean 'cold' as in temperature? Or 'cold' as in cough? Is this person's stomach 'killing them' literally? Or just hurting a bit? Point is, medicine is really, really complicated, and every single layer of complexity makes the data a little less penetrable for a machine.[46]

IBM aren't the only blue-chip big boys to have struggled with the messy, unstructured problem of healthcare data. In 2016, DeepMind, the artificial intelligence arm of Google, signed a contract with the Royal Free NHS Trust in London. DeepMind was granted access to the medical data from three of the city's hospitals in return for an app that could help doctors identify acute kidney injuries. The initial intention was to use clever learning algorithms to help with healthcare; but the researchers found that they had to rein in their ambitions and opt for something much simpler, because the data just wasn't good enough for them to reach their original goals.

Beyond these purely practical challenges, DeepMind's collaboration with the NHS raised a more controversial issue. The researchers only ever promised to alert doctors to kidney injuries, but the Royal Free didn't have a kidney dataset to give them. So instead DeepMind was granted access to *everything* on record: medical histories for some 1.6 million patients going back over a full five years.

In theory, having this incredible wealth of information could help to save innumerable lives. Acute kidney injuries kill one thousand people a month, and having data that reached so far back could potentially help DeepMind to identify important historical trends. Plus, since kidney injuries are more common among people with other diseases, a broad dataset would make it much easier to hunt for clues and connections to people's future health.

Instead of excitement, though, news of the project was met with outrage. And not without justification. Giving DeepMind access to everything on record meant exactly that. The company was told who was admitted to hospital and when. Who came to visit patients during their stay. The results of pathology reports, of radiology exams. Who'd had abortions, who'd had depression, even who had been diagnosed with HIV. And worst of all? The patients themselves were never asked for their consent, never given an opt-out, never even told they were to be part of the study.[47]

It's worth adding that Google was forbidden to use the information in any other part of its business. And – in fairness – it does have a much better track record on data security than the NHS, whose hospitals were brought to a standstill by a North Korean ransomware computer virus in 2017 because it was still running Windows XP.[48] But even so, there is something rather troubling about an already incredibly powerful, world-leading technology company having access to that kind of information about you as an individual.

Problems with privacy

Let's be honest, Google isn't exactly short of private, even intimate information on each of us. But something feels instinctively different – especially confidential – about our medical records.

For anyone with a clean bill of health, it might not be immediately obvious why that should be: after all, if you had to choose between releasing your medical data and your browser history to the world, which would you prefer? I know I'd pick the former in a heartbeat and suspect that many others would too. It's not that I have anything particularly interesting to hide. But one is just a bland snapshot of my biological self, while the other is a window straight into my character.

Even if healthcare data might have less potential to be embarrassing, Timandra Harkness, author of *Big Data: Does Size Matter?* and presenter of the BBC Radio 4 programme *Future Proofing*, argues that it is still a special case.

'First off, a lot of people's healthcare data includes a narrative of their life,' she told me. 'For instance, one in three women in Britain have had a termination – there might be people in their own lives who don't know that.' She also points out that your medical record isn't just relevant to you. 'If someone has your genetic data, they also know something about your parents, your siblings, your children.' And once it's out, there's no getting away from it. 'You cannot change your biology, or deny it. If somebody samples your DNA you can't change that. You can have plastic surgery on your face, you can wear gloves to cover your fingerprints, but your DNA is always there. It's always linked to you.'

Why does this matter? Timandra told me about a focus group she'd chaired in 2013, where ordinary people were asked what concerned them most about their medical data. 'On the whole, people weren't that worried about their data being hacked or stolen. They were concerned about having assumptions made about them as a group and then projected on to them as individuals.'

Most of all, they were concerned with how their data might be used against them. 'Suppose someone had linked their supermarket loyalty card to their medical records. They might go for a hip operation, the doctor would say, 'Oh, I'm sorry, we see here that you've been buying a lot of pizzas or you've been buying a lot of cigarettes and therefore I'm afraid we're going to have to send you to the back of the waiting list.'

That's a very sensible fear in the UK, where some cash-strapped NHS hospitals are already prioritizing non-smokers for knee and hip operations.[49] And there are many countries around the world where insurance or treatment can be denied to obese patients.[50]

There's something of a conundrum here. Humans, as a species, could stand to benefit enormously from having our medical records opened up to algorithms. Watson doesn't have to remain a fantasy. But to turn it into reality, we'll need to hand over our records to companies rich enough to drag us through the slog of the challenges that lie between us and that magical electronic doctor. And in giving up our privacy we'll always be dancing with the danger that our records could be compromised, stolen or used against us. Are you prepared to take that risk? Do you believe in these algorithms and their benefits enough to sacrifice your privacy?

Or, if it comes to it, will you even particularly care?

Genetic giveaway

Francis Galton was a Victorian statistician, a human geneticist, one of the most remarkable men of his generation – and the half-cousin of Charles Darwin. Many of his ideas had a profound effect on modern science, not least his work that essentially laid the foundations for modern statistics. For that, we owe Galton a sincere debt of gratitude. (Unfortunately, he was also active in the burgeoning eugenics movement, for which we certainly do not.)

Galton wanted to study human characteristics through data, and he knew even then that you needed a lot of it to really learn anything of interest. But he also knew that people had an insatiable curiosity about their own bodies. He realized that – when whetted in the right way – people's voracious appetite for an expert's assessment of themselves could over-ride their desire for privacy. What's more, they were often willing to pay for the pleasure of feeding it.

And so, in 1884, when a huge exhibition was held in London under the patronage of Queen Victoria to celebrate the advances Britain had made in healthcare, Galton saw his opportunity. At his own expense, he set up a stand at the exhibition – he called it an 'Anthropometric Laboratory' – in the hopes of finding a few people among the millions of visitors who'd want to pay money to be measured.

He found more than a few. Punters queued up outside and thronged at the door, eager to hand over a threepenny bit each to enter the lab. Once inside, they could pit their skills against a series of specially designed instruments, testing, among other things, their keenness of sight, judgement of eye, strength of pull and squeeze, and swiftness of blow.[51] Galton's lab was so popular he had to take two people through at a time. (He quickly noticed that it was best to keep parents and children apart during the testing to avoid any wasted time on bruised egos. Writing in a journal article after the event, he commented: 'The old did not like to be outdone by the young and insisted on repeated trials.')[52]

However well or badly each person performed, their results were scribbled on to a white card that was theirs to keep as a souvenir. But the real winner was Galton. He left the exhibition with a full copy of everything – a valuable set of biometric measurements for 9,337 people – and a handsome profit to boot.

Fast-forward 130 years and you might just recognize some similarities in the current trend for genetic testing. For the bargain price of £149, you can send off a saliva sample to the genomics and biotechnology company 23andMe in exchange for a report of your genetic traits, including answers to questions such as: What kind of earwax do you have? Do *you* have the genes for a monobrow?[53] Or the gene that makes you sneeze when you look at the sun?[54] As well as some more serious ones: Are you prone to breast cancer? Do you have a genetic predisposition for Alzheimer's disease?[55]

The company, meanwhile, has cleverly amassed a gigantic database of genetic information, now stretching into the millions of samples. It's just what the internet giants have been doing, except in handing over our DNA as part of the trade, we're offering up the most personal data we have. The result is a database we all stand to benefit from. It constitutes a staggeringly valuable asset in terms of its potential to advance our understanding of the human genome. Academics, pharmaceutical companies and non-profits around the world are queuing up to partner with 23andMe to hunt for patterns in their data – both with and without the help of algorithms – in the hope of answering big questions that affect all of us: What are the hereditary causes of different diseases? Are there new drugs that could be invented to treat people with particular conditions? Is there a better way to treat Parkinson's?

The dataset is also valuable in a much more literal sense. Although the research being done offers an immense benefit to society, 23andMe isn't doing this out of the goodness of its heart. If you give it your consent (and 80 per cent of customers do), it will sell on an anonymized version of your genetic data to those aforementioned research partners for a tidy profit.[56] The money earned isn't a happy bonus for the company; it's actually their business plan. One 23andMe board member told *Fast Company*: 'The long game here is not to make money selling kits, although the kits are essential to get the base level data.' Something worth remembering whenever you send off for a commercial genetic report: you're not using the product; you *are* the product.[57]

Word of warning. I'd also be a bit wary of those promises of anonymity. In 2005 a young man,[58] conceived through a completely anonymous sperm donor, managed to track down and identify his birth father by sending off a saliva swab to be analysed and picking up on clues in his own DNA code.[59] Then in 2013 a group of academics, in a particularly famous paper, demonstrated that millions of people could potentially be identified by their genes using nothing more than a home computer and a few clever internet searches.[60]

And there's another reason why you might not want your DNA in any dataset. While there are laws in place to protect people from the worst kinds of genetic discrimination – so we're not quite headed for a future where a Beethoven or a Stephen Hawking will be judged by their genetic predispositions rather than their talent – these rules don't apply to life insurance. No one can make you take a DNA test if you don't want to, but in the US, insurers can ask you if you've already taken a test that calculates your risk of developing a particular disease such as Parkinson's, Alzheimer's or breast cancer and deny you life insurance if the answer isn't to their liking. And in the UK insurers are allowed to take the results of genetic tests for Huntington's disease into account (if the cover is over £500,000).[61] You could try lying, of course, and pretend you've never had the test, but doing so would invalidate your policy. The only way to avoid this kind of discrimination is never to have the test in the first place. Sometimes, ignorance really can be bliss.

The fact is, there's no dataset that would be more valuable in understanding our health than the sequenced genomes of millions of people. But, none the less, I probably won't be getting a genetic test any time soon. And yet (thankfully for society) millions of people *are* voluntarily giving up their data. At the last count 23andMe has more than 2 million genotyped customers,[62] while MyHeritage, Ancestry.com – even the National Geographic Genographic project – have millions more. So perhaps this is a conundrum that isn't. After all, the market has spoken: in a straight swap for your privacy, the chance to contribute to a great public good might not be worth it, but finding out you're 25 per cent Viking certainly is.*

The greatest good?

OK, I'm being facetious. No one can reasonably be expected to keep the grand challenges of the future of human healthcare in the forefront of their minds when deciding whether to send off for a genetic test. Indeed, it's perfectly understandable that people don't – we have different sets of priorities for ourselves as individuals and for humankind as a whole.

But this does bring us to a final important point. If a diagnostic machine capable of recommending treatments *can* be built, who

---

* You can't actually tell if someone is Viking or not, as my good friend the geneticist Adam Rutherford has informed me at length. I mostly put this in to wind him up. To understand the actual science behind why, read his book *A Brief History of Everyone Who Ever Lived: The Stories in Our Genes* (London: Weidenfeld & Nicolson, 2016).

should it serve? The individual or the population? Because there will be times where it may have to choose.

Imagine, for instance, you go into the doctor with a particularly annoying cough. You'll probably get better on your own, but if a machine were serving you – the patient – it might want to send you for an X-ray and a blood test, just to be on the safe side. And it would probably give you antibiotics too, if you asked for them. Even if they only shortened your suffering by a few days, if the sole objective was your health and comfort, the algorithm might decide it was worth the prescription.

But if a machine were built to serve the entire population, it would be far more mindful of the issues around antibiotic resistance. As long as you weren't in immediate danger, your temporary discomfort would pale into insignificance and the algorithm would only dole out the drugs when absolutely necessary. An algorithm like this might also be wary of wasting resources or conscious of long waiting lists, and so not send you for further tests unless you had other symptoms of something more serious. Frankly, it would probably tell you to take some aspirin and stop being such a wimp.

Likewise, a machine working on behalf of everyone might prioritize 'saving as many lives as possible' as its core objective when deciding who should get an organ transplant. Which might well produce a different treatment plan from a machine that only had your interests in mind.

A machine working for the NHS or an insurer might try to minimize costs wherever possible, while one designed to serve a pharmaceutical company might aim to promote the use of one particular drug rather than another.

The case of medicine is certainly less fraught with tension than the examples from criminal justice. There is no defence and prosecution here. Everyone in the healthcare system is working towards the same goal – getting the patient better. But even here every party in the process has a subtly different set of objectives.

In whatever facet of life an algorithm is introduced, there will always be some kind of a balance. Between privacy and public good. Between the individual and the population. Between different challenges and priorities. It isn't easy to find a path through the tangle of incentives, even when the clear prize of better healthcare for all is at the end.

But it's even harder when the competing incentives are hidden from view. When the benefits of an algorithm are over-stated and the risks are obscured. When you have to ask yourself what you're being told to believe, and who stands to profit from you believing it.

Cars

THE SUN WAS BARELY UP above the horizon first thing in the morning of 13 March 2004, but the Slash X saloon bar, in the middle of the Mojave desert, was already thronging with people.[1] The bar is on the outskirts of Barstow, a small town between Los Angeles and Las Vegas, near where Uma Thurman was filmed crawling out of a coffin for *Kill Bill II*.[2] It's a place popular with cowboys and off-roaders, but on that spring day it had drawn the attention of another kind of crowd. The makeshift stadium that had been built outside in the dust was packed with crazy engineers, excited spectators and foolhardy petrolheads who all shared a similar dream. To be the first people on earth to witness a driverless car win a race.

The race had been organized by the US Defence Advanced Research Projects Agency, DARPA (nicknamed the Pentagon's 'mad science' division).[3] The agency had been interested in unmanned vehicles for a while, and with good reason: roadside bombs and targeted attacks on military vehicles were a big cause of deaths on the battlefield. Earlier that year, they had announced their intention to make one-third of US ground military forces vehicles autonomous by 2015.[4]

Up to that point, progress had been slow and expensive. DARPA had spent around half a billion dollars over two decades funding research work at universities and companies in the hope of achieving their ambition.[5] But then they had an ingenious idea: why not create a competition? They would openly invite any interested people across the country to design their own driverless cars and race them against each other on a long-distance track, with a prize of $1 million for the winner.[6] It would be the first event of its kind in the world, and a quick and cheap way to give DARPA a head start in pursuing their goal.

The course was laid out over 142 miles, and DARPA hadn't made it easy. There were steep climbs, boulders, dips, gullies, rough terrain and the odd cactus to contend with. The driverless cars would have to navigate dirt tracks that were sometimes only a few feet wide. Two hours before the start, the organizers gave each team a CD of GPS coordinates.[7] These represented two thousand waypoints that were sprinkled along the route like breadcrumbs – just enough to give the cars a rough sense of where to go, but not enough to help them navigate the obstacles that lay ahead.

The challenge was daunting, but 106 plucky teams applied in that first year. Fifteen competitors passed the qualifying rounds and were considered safe enough to take on the track. Among them were cars that looked like dune buggies, cars that looked like monster trucks, and cars that looked like tanks. There were rumours that one contender had mortgaged their house to build their car, while another had two surfboards stuck on their vehicle roof to make it stand out. There was even a self-balancing motorbike.[8]

On the morning of the race, a ramshackle line-up of cars gathered at Slash X along with a few thousand spectators. Without any drivers inside, the vehicles took turns approaching the start, each one looking more like it belonged in *Mad Max* or *Wacky Races* than the last. But looks didn't matter. All they had to do was get around the track, without any human intervention, in less than ten hours.

Things didn't quite go as planned. One car flipped upside down in the starting area and had to be withdrawn.[9] The motorbike barely cleared the start line before it rolled on to its side and was declared out of the race. One car hit a concrete wall 50 yards in. Another got tangled in a barbed-wire fence. Yet another got caught between two tumbleweeds and – thinking they were immovable objects – became trapped reversing back and forth, back and forth, until someone eventually intervened.[10] Others still went crashing into boulders and careering into ditches. Axles were snapped, tyres ripped, bodywork went flying.[11] The scene around the Slash X saloon bar began to look like a robot graveyard.

The top-scoring vehicle, an entry by Carnegie Mellon University, managed an impressive 7 miles before misjudging a hill – at which point the tyres started spinning and, without a human to help, carried on spinning until they caught fire.[12] By 11 a.m. it was all over. A DARPA organizer climbed into a helicopter and flew over to the finish line to inform the waiting journalists that none of the cars would be getting that far.[13]

The race had been oily, dusty, noisy and destructive – and had ended without a winner. All those teams of people had worked for a year on a creation that had lasted at best a few minutes. But the competition was anything but a disaster. The rivalry had led to an explosion of new ideas, and by the next Grand Challenge in 2005, the technology was barely recognizable.

Second time around, all but one of the entrants surpassed the 7 miles achieved in 2004. An astonishing five different cars managed to complete the full race distance of 132 miles without any human intervention.[14]

Now, little more than a decade later, it's widely accepted that the future of transportation is driverless. In late 2017, Philip Hammond, the British Chancellor of the Exchequer, announced the government's intention to have fully driverless cars – without a safety attendant on board – on British roads by 2021. Daimler has promised driverless cars by 2020,[15] Ford by 2021,[16] and other manufacturers have made their own, similar forecasts.

Talk in the press has moved on from questioning whether driverless cars will happen to addressing the challenges we'll face when they do. 'Should your driverless car hit a pedestrian to save your life?' asked the *New York Times* in June 2016;[17] and, in November 2017: 'What happens to roadkill or traffic tickets when our vehicles are in control?'[18] Meanwhile, in January 2018, the *Financial Times* warned: 'Trucks headed for a driverless future: unions warn that millions of drivers' jobs will be disrupted.'[19]

So what changed? How did this technology go from ramshackle incompetence to revolutionary confidence in a few short years? And can we reasonably expect the rapid progress to continue?

What's around me?

Our dream of a perfect autonomous vehicle dates all the way back to the sci-fi era of jet packs, rocket ships, tin-foil space suits and ray guns. At the 1939 World Fair in New York, General Motors unveiled its vision of the future. Visitors to the exhibition strapped themselves into an audio-equipped chair mounted on a conveyor that took them on a 16-minute tour of an imagined world.[20] Beneath the glass, they saw a scale model of the GM dream. Superhighways that spanned the length and breadth of the country, roads connecting farmlands and cities, lanes and intersections – and roaming over all of them, automated radio-controlled cars capable of safely travelling at speeds of up to 100 miles an hour. 'Strange?' the voiceover asked them. 'Fantastic? Unbelievable? Remember, this is the world of 1960!'[21]

There were numerous attempts over the years to make the dream a reality. General Motors tried with the Firebird II in the 1950s.[22] British researchers tried adapting a Citroën DS19 to communicate with the road in the 1960s (somewhere between Slough and Reading, you'll still find a 9-mile stretch of electric cable, left over from their experiments).[23] Carnegie's 'Navlab' in the 1980s, the EU's $1 billion Eureka Prometheus Project in the 1990s.[24] With every new project, the dream of the driverless car seemed, tantalizingly, only just around the corner.

On the surface, building a driverless car sounds as if it should be relatively easy. Most humans manage to master the requisite skills to drive. Plus, there are only two possible outputs: speed and direction. It's a question of how much gas to apply and how much to turn the wheel. How hard can it be?

But, as the first DARPA Grand Challenge demonstrated, building an autonomous vehicle is actually a lot trickier than it looks. Things quickly get complicated when you're trying to get an algorithm to control a great big hunk of metal travelling at 60 miles per hour.

Take the neural networks that are used to great effect to detect tumours in breast tissue; you'd think they should be perfectly suited to help a driverless car technology 'see' its surroundings. By 2004, neural networks (albeit in slightly more rudimentary form than today's state-of-the-art versions) were already whirring away within prototype driverless vehicles,[25] trying to extract meaning from the cameras mounted on top of the cars. There's certainly a great deal of valuable information to be had from a camera. A neural network can understand the colour, texture, even physical features of the scene ahead – things like lines, curves, edges and angles. The question is: what do you do with that information once you have it?

You could tell the car: 'Only drive on something that looks like tarmac.' But that won't be much good in the desert, where the roads are dusty paths. You could say: 'Drive on the smoothest thing in the image' – but, unfortunately, the smoothest thing is almost always the sky or a glass-fronted building. You could think in quite abstract terms about how to describe the shape of a road: 'Look for an object with two vaguely straight borders. The lines should be wide apart at the bottom of the image and taper in towards each other at the top.' That seems pretty sensible. Except, unfortunately, it's also how a tree looks in a photograph. Generally, it isn't considered wise to encourage a car to drive up a tree.

The issue is that cameras can't give you a sense of scale or distance. It's something film directors use to their advantage all the time – think of the opening scene in *Star Wars* where the Star Destroyer slowly emerges against the inky blackness of space, looming dramatically over the top of the frame. You get the sense of it being a vast, enormous beast, when in reality it was filmed using a model no more than a few feet long. It's a trick that works well on the big screen. But in a driverless car, when two thin parallel lines could either be a road ahead on the horizon or the trunk of a nearby tree, accurately judging distance becomes a matter of life and death.

Even if you use more than one camera and cleverly combine the images to build a 3D picture of the world around you, there's another potential problem that comes from relying too heavily on neural networks, as Dean Pomerleau, an academic from Carnegie Mellon University, discovered back in the 1990s. He was working on a car called ALVINN, Autonomous Land Vehicle In a Neural Network, which was trained in how to understand its surroundings from the actions of a human driver. Pomerleau and others would sit at the wheel and take ALVINN on long drives, recording everything they were doing in the process. This formed the training dataset from which their neural networks would learn: drive anywhere a human would, avoid everywhere else.[26]

It worked brilliantly at first. After training, ALVINN was able to comfortably navigate a simple road on its own. But then ALVINN came across a bridge and it all went wrong. Suddenly, the car swerved dangerously, and Pomerleau had to grab hold of the wheel to save it from crashing.

After weeks of going through the data from the incident, Pomerleau worked out what the issue had been: the roads that ALVINN had been trained on had all had grass running along the sides. Just like those neural networks back in the 'Medicine' chapter, which classified huskies on the basis of snow in the pictures, ALVINN's neural network had used the grass as a key indicator of where to drive. As soon as the grass was gone, the machine had no idea what to do.

Unlike cameras, lasers *can* measure distance. Vehicles that use a system called LiDAR (Light Detection and Ranging, first used at the second DARPA Grand Challenge in 2005) fire out a photon from a laser, time how long it takes to bounce off an obstacle and come back, and end up with a good estimate of how far away that obstacle is. It's not all good news: LiDAR can't help with texture or colour, it's hopeless at reading road signs, and it's not great over long distances. Radar, on the other hand – the same idea but with radio waves – does a good job in all sorts of weather conditions, can detect obstacles far away, even seeing through some materials, but is completely hopeless at giving any sort of detail of the shape or structure of the obstacle.

On its own, none of these data sources – the camera, the LiDAR, the radar – can do enough to understand what's going on around a vehicle. The trick to successfully building a driverless car is combining them. Which would be a relatively easy task if they all agreed about what they were actually seeing, but is a great deal more difficult when they don't.

Consider the tumbleweed that stumped one of the cars in the first DARPA Grand Challenge and imagine your driverless car finds itself in the same position. The LiDAR is telling you there is an obstacle ahead. The camera agrees. The radar, which can pass through the flimsy tumbleweed, is telling you there's nothing to worry about. Which sensor should your algorithm trust?

What if the camera pulls rank? Imagine a big white truck crosses your path on a cloudy day. This time LiDAR and radar agree that the brakes need to be applied, but against the dull white sky, the camera can see nothing that represents a danger.

If that weren't hard enough, there's another problem. You don't just need to worry about your sensors misinterpreting their surroundings, you need to take into account that they might mismeasure them too.

You may have noticed that blue circle on Google Maps that surrounds your location – it's there to indicate the potential error in the GPS reading. Sometimes the blue circle will be small and accurately mark your position; at other times it will cover a much larger area and be centred on entirely the wrong place. Most of the time, it doesn't much matter. We know where we are and can dismiss incorrect information. But a driverless car doesn't have a ground truth of its position. When it's driving down a single lane of a motorway, less than 4 metres wide, it can't rely on GPS alone for an accurate enough diagnosis of where it is.

GPS isn't the only reading that's prone to uncertainty. Every measurement taken by the car will have some margin of error: radar readings, the pitch, the roll, the rotations of the wheels, the inertia of the vehicle. Nothing is ever 100 per cent reliable. Plus, different conditions make things worse: rain affects LiDAR;[27] glaring sunlight can affect the cameras;[28] and long, bumpy drives wreak havoc with accelerometers.[29]

In the end, you're left with a big mess of signals. Questions that seemed simple – Where are you? What's around you? What should you do? – become staggeringly difficult to answer. It's almost impossible to know what to believe.

Almost impossible. But not quite.

Because, thankfully, there is a route through all of this chaos – a way to make sensible guesses in a messy world. It all comes down to a phenomenally powerful mathematical formula, known as Bayes' theorem.

The great Church of the Reverend Bayes

It's no exaggeration to say that Bayes' theorem is one of the most influential ideas in history. Among scientists, machine-learning experts and statisticians, it commands an almost cultish enthusiasm. Yet at its heart the idea is extraordinarily simple. So simple, in fact, that you might initially think it's just stating the obvious.

Let me try and illustrate the idea with a particularly trivial example.

Imagine you're sitting having dinner in a restaurant. At some point during the meal, your companion leans over and whispers that they've spotted Lady Gaga eating at the table opposite.

Before having a look for yourself, you'll no doubt have some sense of how much you believe your friend's theory. You'll take into account all of your prior knowledge: perhaps the quality of the establishment, the distance you are from Gaga's home in Malibu, your friend's eyesight. That sort of thing. If pushed, it's a belief that you could put a number on. A probability of sorts.

As you turn to look at the woman, you'll automatically use each piece of evidence in front of you to update your belief in your friend's hypothesis. Perhaps the platinum-blonde hair is consistent with what you would expect from Gaga, so your belief goes up. But the fact that she's sitting on her own with no bodyguards isn't, so your belief goes down. The point is, each new observation adds to your overall assessment.

This is all Bayes' theorem does: offers a systematic way to update your belief in a hypothesis on the basis of the evidence.[30] It accepts that you can't ever be completely certain about the theory you're considering, but allows you to make a best guess from the information available. So, once you realize the woman at the table opposite is wearing a dress made of meat – a fashion choice that you're unlikely to chance upon in the non-Gaga population – that might be enough to tip your belief over the threshold and lead you to conclude that it is indeed Lady Gaga in the restaurant.

But Bayes' theorem isn't just an equation for the way humans already make decisions. It's much more important than that. To quote Sharon Bertsch McGrayne, author of *The Theory That Would Not Die*: 'Bayes runs counter to the deeply held conviction that modern science requires objectivity and precision.'[31] By providing a mechanism to measure your belief in something, Bayes allows you to draw sensible conclusions from sketchy observations, from messy, incomplete and approximate data – even from ignorance.

Bayes isn't there just to confirm our existing intuitions. It turns out that being forced to quantify your beliefs in something often leads to counter-intuitive conclusions. It's Bayes' theorem that explains why more men than women are falsely identified as future murderers, in the example on page 67 in the 'Justice' chapter. And it's Bayes' theorem that explains why – even if you have been diagnosed with breast cancer – the level of error in the tests means you probably don't have it (see the 'Medicine' chapter, page 94). Across all branches of science, Bayes is a powerful tool for distilling and understanding what we really know.

But where the Bayesian way of thinking really comes into its own is when you're trying to consider more than one hypothesis simultaneously – for example, in attempting to diagnose what's wrong with a patient on the basis of their symptoms,* or finding the position of a driverless car on the basis of sensor readings. In theory, any

---

* Watson, the IBM machine discussed in the 'Medicine' chapter, makes extensive use of so-called Bayesian inference. See https://www.ibm.com/developerworks/library/os-ind-watson/.

disease, any point on the map, could represent the underlying truth. All you need to do is weigh up the evidence to decide which is most likely to be right.

And on that point, finding the location of a driverless car turns out to be rather similar to a problem that puzzled Thomas Bayes, the British Presbyterian minister and talented mathematician after whom the theorem is named. Back in the mid-1700s, he wrote an essay which included details of a game he'd devised to explain the problem. It went something a little like this:[32]

Imagine you're sitting with your back to a square table. Without you seeing, I throw a red ball on to the table. Your job is to guess where it landed. It's not going to be easy: with no information to go on, there's no real way of knowing where on the table it could be.

So, to help your guess, I throw a second ball of a different colour on to the same table. Your job is still to determine the location of the first ball, the red one, but this time I'll tell you where the second ball ends up on the table relative to the first: whether it's in front, behind, to the left or right of the red ball. And you get to update your guess.

Then we repeat. I throw a third, a fourth, a fifth ball on to the table, and every time I'll tell you where each one lands relative to the very first red one – the one whose position you're trying to guess.

The more balls I throw and the more information I give you, the clearer the picture of the red ball's position should become in your mind. You'll never be absolutely sure of exactly where it sits, but you can keep updating your belief about its position until you end up with an answer you're confident in.

In some sense, the true position of the driverless car is analogous to that of the red ball. Instead of a person sitting with their back to the table, there's an algorithm trying to gauge exactly where the car is at that moment in time, and instead of the other balls thrown on to the table there are the data sources: the GPS, the inertia measurements and so on. None of them tells the algorithm where the car is, but each adds a little bit more information the algorithm can use to update its belief. It's a trick known as probabilistic inference – using the data (plus Bayes) to infer the true position of the object. Packaged up correctly, it's just another kind of machine-learning algorithm.

By the turn of the millennium, engineers had had enough practice with cruise missiles, rocket ships and aircraft to know how to tackle the position problem. Getting a driverless car to answer the question 'Where am I?' still wasn't trivial, but with a bit of Bayesian thinking it was at least achievable.

Between the robot graveyard of the 2004 Grand Challenge and the awe-inspiring technological triumph of the 2005 event – when five different vehicles managed to race more than 100 miles without any human input – many of the biggest leaps forward were thanks to Bayes. It was algorithms based on Bayesian ideas that helped solve the other questions the car needed to answer: 'What's around me?' and 'What should I do?'*

So, should your driverless car hit a pedestrian to save your life?

Let's pause for a moment to consider the second of those questions. Because, on this very topic, in early autumn 2016, tucked away in a quiet corner of an otherwise bustling exhibition hall at the Paris Auto Show, a Mercedes-Benz spokesperson made a rather exceptional statement. Christoph von Hugo, the manager of driver

---

* The eventual winner of the 2005 race, a team from Stanford University, was described rather neatly by the Stanford University mathematician Pesri Diaconis: 'Every bolt of that car was Bayesian.'

assistance systems and active safety for the company, was asked in an interview what a driverless Mercedes might do in a crash.

'If you know you can save at least one person, at least save that one,' he replied.[33]

Sensible logic, you would think. Hardly headline news.

Except, Hugo wasn't being asked about any old crash. He was being tested on his response to a well-worn thought experiment dating back to the 1960s, involving a very particular kind of collision. The interviewer was asking him about a curious conundrum that forces a choice between two evils. It's known as the trolley problem, after the runaway tram that was the subject of the original formulation. In the case of driverless cars, it goes something like this.

Imagine, some years into the future, you're a passenger in an autonomous vehicle, happily driving along a city street. Ahead of you a traffic light turns red, but a mechanical failure in your car means you're unable to stop. A collision is inevitable, but your car has a choice: should it swerve off the road into a concrete wall, causing certain death to anyone inside the vehicle? Or should it carry on going, saving the lives of anyone inside, but killing the pedestrians now crossing the road? What should the car be programmed to do? How do you decide who should die?

No doubt you have your own opinion. Perhaps you think the car should simply try to save as many lives as possible. Or perhaps you think that 'thou shalt not kill' should over-ride any calculations, leaving the one sitting in the machine to bear the consequences.*

Hugo was clear about the Mercedes position. 'Save the one in the car.' He went on: 'If all you know for sure is that one death can be prevented, then that's your first priority.'

---

* A number of different versions of the scenario have appeared across the press, from the *New York Times* to the *Mail on Sunday*: What if the pedestrian was a 90-year-old granny? What if it was a small child? What if the car contained a Nobel Prize winner? All have the same dilemma at the core.

In the days following the interview, the internet was awash with articles berating Mercedes' stance. 'Their cars will act much like the stereotypical entitled European luxury car driver,'[34] wrote the author of one piece. Indeed, in a survey published in *Science* that very summer,[35] 76 per cent of respondents felt it would be more moral for driverless vehicles to save as many lives as possible, thus killing the people within the car. Mercedes had come down on the wrong side of popular opinion.

Or had they? Because when the same study asked participants if they would actually *buy* a car which would murder them if the circumstances arose, they suddenly seemed reluctant to sacrifice themselves for the greater good.

This is a conundrum that divides opinion – and not just in what people think the answer should be. As a thought experiment, it remains a firm favourite of technology reporters and other journalists, but all the driverless car experts I interviewed rolled their eyes as soon as the trolley problem was mentioned. Personally, I still have a soft spot for it. Its simplicity forces us to recognize something important about driverless cars, to challenge how we feel about an algorithm making a value judgement on our own, and others', lives. At the heart of this new technology – as with almost all algorithms – are questions about power, expectation, control, and delegation of responsibility. And about whether we can expect our technology to fit in with us, rather than the other way around. But I'm also sympathetic to the aloof reaction it receives in the driverless car community. They, more than anyone, know how far away we are from having to worry about the trolley problem as a reality.

Breaking the rules of the road

Bayes' theorem and the power of probability have driven much of the innovation in autonomous vehicles ever since the DARPA challenge. I asked Paul Newman, professor of robotics at the University of Oxford and founder of Oxbotica, a company that builds driverless cars and tests them on the streets of Britain, how his latest autonomous vehicles worked, and he explained as follows: 'It's many, many millions of lines of code, but I could frame the entire thing as probabilistic inference. All of it.'[36]

But while Bayesian inference goes some way towards explaining how driverless cars are possible, it also explains how full autonomy, free from any input by a human driver, is a very, very difficult nut to crack.

Imagine, Paul Newman suggests, 'you've got two vehicles approaching each other at speed' – say, travelling in different directions down a gently curved A-road. A human driver will be perfectly comfortable in that scenario, knowing that the other car will stick to its own lane and pass safely a couple of metres to the side. 'But for the longest time,' Newman explains, 'it does look like you're going to hit each other.' How do you teach a driverless car not to panic in that situation? You don't want the vehicle to drive off the side of the road, trying to avoid a collision that was never going to happen. But, equally, you don't want it to be complacent if you really do find yourself on the verge of a head-on-crash. Remember, too, these cars are only ever making educated guesses about what to do. How do you get it to guess right every single time? That, says Newman, 'is a hard, hard problem'.

It's a problem that puzzled the experts for a long time, but it does have a solution. The trick is to build in a model for how other – sane – drivers will behave. Unfortunately, the same can't be said of other nuanced driving scenarios.

Newman explains: 'What's hard is all the problems with driving that have nothing to do with driving.' For instance, teaching an algorithm to understand that hearing the tunes of an ice-cream van, or passing a group of kids playing with a ball on the pavement, might mean you need to be extra cautious. Or to recognize the confusing hopping of a kangaroo, which, at the time of writing, Volvo admitted to be struggling with.[37] Probably not much of a problem in rural Surrey, but something the cars need to master if they're to be roadworthy in Australia.

Even harder, how do you teach a car that it should sometimes break the rules of the road? What if you're sitting at a red light and someone runs in front of your car and frantically beckons you to edge forwards? Or if an ambulance with its lights on is trying to get past on a narrow street and you need to mount the pavement to let it through? Or if an oil tanker has jack-knifed across a country lane and you need to get out of there by any means possible?

'None of these are in the Highway Code,' Newman rightly points out. And yet a truly autonomous car needs to know how to deal with all of them if it's to exist without *ever* having any human intervention. Even in emergencies.

That's not to say these are insurmountable problems. 'I don't believe there's any level of intelligence that we won't be able to get a machine to,' Newman told me. 'The only question is when.'

Unfortunately, the answer to that question is: probably not any time soon. That driverless dream we're all waiting for might be quite a lot further away than we think.

Because there's another layer of difficulty to contend with when trying to build that sci-fi fantasy of a go-anywhere, do-anything, steering-wheel-free driverless car, and it's one that goes well beyond the technical challenge. A fully autonomous car will also have to deal with the tricky problem of people.

Jack Stilgoe, a sociologist from University College London and an expert in the social impact of technology, explains: 'People are mischievous. They're active agents, not just passive parts of the scenery.'[38]

Imagine, for a moment, a world where truly, perfectly autonomous vehicles exist. The number one rule in their on-board algorithms will be to avoid collisions wherever possible. And that changes the dynamics of the road. If you stand in front of a driverless car – it has to stop. If you pull out in front of one at a junction – it has to behave submissively.

In the words of one participant in a 2016 focus group at the London School of Economics: 'You're going to mug them right off. They're going to stop and you're just going to nip round.' Translation: these cars can be bullied.

Stilgoe agrees: 'People who've been relatively powerless on roads up 'til now, like cyclists, may start cycling very slowly in front of self-driving cars, knowing that there is never going to be any aggression.'

Getting around this problem might mean bringing in stricter rules to deal with people who abuse their position as cyclists or pedestrians. It's been done before, of course: think of jay-walking. Or it could mean forcing everything else off the roads – as happened with the introduction of the motor car – which is why already you won't see bicycles, horses, carts, carriages or pedestrians on a motorway.

If we want fully autonomous cars, we'll almost certainly have to do something similar again and limit the number of aggressive drivers, ice-cream vans, kids playing in the road, roadwork signs, difficult pedestrians, emergency vehicles, cyclists, mobility scooters and everything else that makes the problem of autonomy so difficult. That's fine, but it's a little different from the way the idea is currently being sold to us.

'The rhetoric of autonomy and transport is all about *not* changing the world,' Stilgoe tells me. 'It's about keeping the world as it is but making and allowing a robot to just be as good as and then better than a human at navigating it. And I think that's stupid.'

But hang on, some of you may be thinking. Hasn't this problem already been cracked? Hasn't Waymo, Google's autonomous car, driven millions of miles already? Aren't Waymo's fully autonomous cars (or at least, close to fully autonomous cars) currently driving around on the roads of Phoenix, Arizona?

Well, yes. That's true. But not every mile of road is created equally. Most miles are so easy to drive, you can do it while daydreaming. Others are far more challenging. At the time of writing, Waymo cars aren't allowed to go just anywhere: they're 'geo-fenced' into a small, pre-defined area. So too are the driverless cars Daimler and Ford propose to have on the roads by 2020 and 2021 respectively. They're ride-hailing cars confined to a pre-decided go-zone. And that does make the problem of autonomy quite a lot simpler.

Paul Newman thinks this is the future of driverless cars we can expect: 'They'll come out working in an area that's very well known, where their owners are extremely confident that they'll work. So it could be part of a city, not in the middle of a place with unusual roads or where cows could wander into the path. Maybe they'll work at certain times of day and in certain weather situations. They're going to be operated as a transport service.'

That's not quite the same thing as full autonomy. Here's Jack Stilgoe's take on the necessary compromise: 'Things that look like autonomous systems are actually systems in which the world is constrained to make them look autonomous.'

The vision we've come to believe in is like a trick of the light. A mirage that promises a luxurious private chauffeur for all but, close up, is actually just a local minibus.

If you still need persuading, I'll leave the final word on the matter to one of America's biggest automotive magazines – *Car and Driver*:

> No car company actually expects the futuristic, crash-free utopia of streets packed with driverless vehicles to transpire anytime soon, nor for decades. But they do want to be taken seriously by Wall Street as well as stir up the imaginations of a public

increasingly disinterested in driving. And in the meantime, they hope to sell lots of vehicles with the latest sophisticated driver-assistance technology.[39]

So how about that driver-assistance technology? After all, driverless cars are not an all-or-nothing proposition.

Driverless technology is categorized using six different levels: from level 0 – no automation whatsoever – up to to level 5 – the fully autonomous fantasy. In between, they range from cruise control (level 2) to geo-fenced autonomous vehicles (level 4) and are colloquially referred to as level 1: feet off; level 2: hands off; level 3: eyes off; level 4: brain off.

So, maybe level 5 isn't on our immediate horizon, and level 4 won't be quite what it's cracked up to be, but there's a whole lot of automation to be had on the way up. What's wrong with just slowly working up the levels in our private cars? Build cars with steering wheels and brake pedals and drivers in driver's seats, and just allow a human to step in and take over in an emergency? Surely that'll do until the technology improves?

Unfortunately, things aren't quite that simple. Because there's one last twist in the tale. A whole host of other problems. An inevitable obstacle for anything short of total human-free driving.

The company baby

Among the pilots at Air France, Pierre-Cédric Bonin was known as a 'company baby'.[40] He had joined the airline at the tender age of 26, with only a few hundred hours of flying time under his belt, and had grown up in the airline's fleet of Airbuses. By the time he stepped aboard the fated flight of AF447, aged 32, he had managed to clock up a respectable 2,936 hours in the air, although that still made him by far the least experienced of the three pilots on board.[41]

None the less, it was Bonin who sat at the controls of Air France flight 447 on 31 May 2009, as took it off from the tarmac of Rio de Janeiro–Galeão International Airport and headed home to Paris.[42]

This was an Airbus A330, one of the most sophisticated commercial aircraft ever built. Its autopilot system was so advanced that it was practically capable of completing an entire flight unaided, apart from take-off and landing. And even when the pilot was in control, it had a variety of built-in safety features to minimize the risk of human error.

But there's a hidden danger in building an automated system that can safely handle virtually every issue its designers can anticipate. If a pilot is only expected to take over in exceptional circumstances, they'll no longer maintain the skills they need to operate the system themselves. So they'll have very little experience to draw on to meet the challenge of an unanticipated emergency.

And that's what happened with Air France flight 447. Although Bonin had accumulated thousands of hours in an Airbus cockpit, his actual experience of flying an A330 by hand was minimal. His role as a pilot had mostly been to monitor the automatic system. It meant that when the autopilot disengaged during that evening's flight, Bonin didn't know how to fly the plane safely.[43]

The trouble started when ice crystals began to form inside the air-speed sensors built into the fuselage. Unable to take a sensible reading, the autopilot sounded an alarm in the cabin and passed responsibility to the human crew. This in itself was not cause for concern. But when the plane hit a small bump of turbulence, the inexperienced Bonin over-reacted. As the aircraft began to roll gently to the right, Bonin grabbed the side-stick and pulled it to the left. Crucially, at the same time, he pulled back on the stick, sending the aircraft into a dramatically steep climb.[44]

As the air thinned around the plane, Bonin kept tightly pulling back on the stick until the nose of the aircraft was so high that the air could no longer flow slickly over the wings. The wings effectively became wind-breakers and the aircraft dramatically lost lift, free-falling, nose-up, out of the sky.

Alarms sounded in the cockpit. The captain burst back in from the rest cabin. AF447 was descending towards the ocean at 10,000 feet per minute.

By now, the ice crystals had melted, there was no mechanical malfunction, and the ocean was far enough below them that they could still recover in time. Bonin and his co-pilot could have easily rescued everyone on board in just 10–15 seconds, simply by pushing the stick forward, dropping the aircraft's nose and allowing the air to rush cleanly over the wings again.[45]

But in his panic, Bonin kept the side-stick pulled back. No one realized he was the one causing the issue. Precious seconds ticked by. The captain suggested levelling the wings. They briefly discussed whether they were ascending or descending. Then, within 8,000 feet of sea level, the co-pilot took the controls.[46]

'Climb … climb … climb … climb …' the co-pilot is heard shouting.

'But I've had the stick back the whole time!' Bonin replied.

The penny dropped for the captain. He finally realized they had been free-falling in an aerodynamic stall for more than three minutes and ordered them to drop the nose. Too late. Tragically, by now, they were too close to the surface. Bonin screamed: 'Damn it! We're going to crash. This can't be happening!'[47] Moments later the aircraft plunged into the Atlantic, killing all 228 souls on board.

Ironies of automation

Twenty-six years before the Air France crash, in 1983, the psychologist Lisanne Bainbridge wrote a seminal essay on the hidden dangers of relying too heavily on automated systems.[48] Build a machine to improve human performance, she explained, and it will lead – ironically – to a reduction in human ability.

By now, we've all borne witness to this in some small way. It's why people can't remember phone numbers any more, why many of us struggle to read our own handwriting and why lots of us can't navigate anywhere without GPS. With technology to do it all for us, there's little opportunity to practise our skills.

There is some concern that the same might happen with self-driving cars – where the stakes are a lot higher than with handwriting. Until we get to full autonomy, the car will still sometimes unexpectedly hand back control to the driver. Will we be able to remember instinctively what to do? And will teenage drivers of the future ever have the chance to master the requisite skills in the first place?

But even if all drivers manage to stay competent* (allowing for generous interpretation of the word 'stay'), there's another issue we'll still have to contend with. Because what the human driver is asked to do before autopilot cuts out is also important. There are only two possibilities. And – as Bainbridge points out – neither is particularly appealing.

A level 2, hands-off car will expect the driver to pay careful attention to the road at all times.[49] It's not good enough to be trusted on its own and will need your careful supervision. *Wired* once described this level as 'like getting a toddler to help you with the dishes'.[50]

At the time of writing, Tesla's autopilot is one such example of this approach.[51] It's currently like a fancy cruise control – it'll steer and brake and accelerate on the motorway, but expects the driver to be alert and attentive and ready to step in at all times. To make sure

---

* There are things you can do to tackle the issues that arise from limited practice. For instance, since the Air France crash, there is now an emphasis on training new pilots to fly the plane when autopilot fails, and on prompting all pilots to regularly switch autopilot off to maintain their skills.

you're paying attention, an alarm will sound if you remove your hands from the wheel for too long.

But, as Bainbridge wrote in her essay, that's not an approach that's going to end well. It's just unrealistic to expect humans to be vigilant: 'It's impossible for even a highly motivated human to maintain effective visual attention towards a source of information, on which very little happens, for more than about half an hour.'[52]

There's some evidence that people have struggled to heed Tesla's insistence that they keep their attention on the road. Joshua Brown, who died at the wheel of his Tesla in 2016, had been using Autopilot mode for 37½ minutes when his car hit a truck that was crossing his lane. The investigation by the National Highway Traffic Safety Administration concluded that Brown had not been looking at the road at the time of the crash.[53] The accident was headline news around the world, but this hasn't stopped some foolhardy YouTubers from posting videos enthusiastically showing how to trick your car into thinking you're paying attention. Supposedly, taping a can of Red Bull[54] or wedging an orange[55] into the steering wheel will stop the car setting off those pesky alarms reminding you of your responsibilities.

Other programmes are finding the same issues. Although Uber's driverless cars need human intervention every 13 miles,[56] getting drivers to pay attention remains a struggle. On 18 March 2018, an Uber self-driving vehicle fatally struck a pedestrian. Video footage from inside the car showed that the 'human monitor' sitting behind the wheel was looking away from the road in the moments before the collision.[57]

This is a serious problem, but there is an alternative option. The car companies could accept that humans will be humans, and acknowledge that our minds will wander. After all, being able to read a book while driving is part of the *appeal* of self-driving cars.

This is the key difference between level 2: 'hands off' and level 3: 'eyes off'.

The latter certainly presents more of a technical challenge than level 2, but some manufacturers have already started to build their cars to accommodate our inattention. Audi's traffic-jam pilot is one such example.[58] It can completely take over when you're in slow-moving highway traffic, leaving you to sit back and enjoy the ride. Just be prepared to step in if something goes wrong.*

There's a reason why Audi has limited its system to slow-moving traffic on limited-access roads. The risks of catastrophe are lower in motorway congestion. And that's important. Because as soon as a human stops monitoring the road, you're left with the worst possible combination of circumstances when an emergency happens.

A driver who's not paying attention will have very little time to assess their surroundings and decide what to do. Imagine sitting in a self-driving car, hearing an alarm and looking up from your book to see a truck ahead shedding its load into your path. In an instant, you'll have to process all the information around you: the motorbike in the left lane, the van braking hard ahead, the car in the blind spot on your right. You'd be most unfamiliar with the road at precisely the moment you need to know it best; add in the lack of practice, and you'll be as poorly equipped as you could be to deal with the situations demanding the highest level of skill.

It's a fact that has also been borne out in experiments with driverless car simulations. One study, which let people read a book or play on their phones while the car drove itself, found that it took up to 40 seconds after an alarm sounded for them to regain proper control

---

* A step up from partial automation, level 3 vehicles like the Audi with the traffic-jam pilot can take control in certain scenarios, if the conditions are right. The driver still needs to be prepared to intervene when the car encounters a scenario it doesn't understand, but no longer needs to continuously monitor the road and car. This level is a bit more like persuading a teenager to do the washing up.

of the vehicle.[59] That's exactly what happened with Air France flight 447. Captain Dubois, who should have been easily capable of saving the plane, took around one minute too long to realize what was happening and come up with the simple solution that would have solved the problem.[60]

Ironically, the better self-driving technology gets, the worse these problems become. A sloppy autopilot that sets off an alarm every 15 minutes will keep a driver continually engaged and in regular practice. It's the smooth and sophisticated automatic systems that are *almost* always reliable that you've got to watch out for.

This is why Gill Pratt, who heads up Toyota's research institute, has said:

> The worst case is a car that will need driver intervention once every 200,000 miles ... An ordinary person who has a [new] car every 100,000 miles would never see it [the automation hand over control]. But every once in a while, maybe once for every two cars that I own, there would be that one time where it suddenly goes 'beep beep beep, now it's your turn!' And the person, typically having not seen this for years and years, would ... not be prepared when that happened.[61]

Great expectations

Despite all this, there's good reason to push ahead into the self-driving future. The good still outweighs the bad. Driving remains one of the biggest causes of avoidable deaths in the world. If the technology is remotely capable of reducing the number of fatalities on the roads overall, you could argue that it would be unethical *not* to roll it out.

And there's no shortage of other advantages: even simple self-driving aids can reduce fuel consumption[62] and ease traffic congestion.[63] Plus – let's be honest – the idea of taking your hands off the steering wheel while doing 70 miles an hour, even if it's only for a moment, is just ... cool.

But, thinking back to Bainbridge's warnings, they do hint at a problem with how current self-driving technology is being framed.

Take Tesla, one of the first car manufacturers to bring an autopilot to the market. There's little doubt that their system has had a net positive impact, making driving safer for those who use it – you don't need to look far to find online videos of the 'Forward Collision Warning' feature, recognizing the risk of accident before the driver, setting off an alarm and saving the car from crashing.[64]

But there's a slight mismatch between what the cars can do – with what's essentially a fancy forward-facing parking sensor and clever cruise control – and the language used to describe them. For instance, in October 2016 the company announced that 'all Tesla cars being produced now have full self-driving hardware'.* According to an article in *The Verge*, Elon Musk, product architect of Tesla, added: 'The full autonomy update will be standard on all Tesla vehicles from here on out.'[65] And that phrase 'full autonomy' is arguably at odds with the warning users must accept before using the current autopilot: 'You need to maintain control and responsibility of your vehicle.'[66]

Expectations are important. You may disagree, but I think that people shoving oranges in their steering wheels – or worse, as I found in the darker corners of the Web, creating and selling devices that '[allow] early adopters to [drive, while] reducing or turning off the autopilot check-in warning'† – is the inevitable corollary of a trusted brand using language that misleads.

---

* At the time of writing, in February 2018, the 'full self-driving hardware' is an optional extra that can be paid for at purchase, although the car is not currently running the software to complete full self-driving trips. The Tesla website says: 'It is not possible to know exactly when each element of the functionality described above will be available.' See https://www.tesla.com/en_GB/blog/all-tesla-cars-being-produced-now-have-full-self-driving-hardware.

† This is a real product called 'The Autopilot Buddy' which you can buy for the bargain price of $179. It's worth noting that the small print on the website reads: 'At no time should 'Autopilot Buddy™' be used on public streets.' https://www.autopilotbuddy.com/.

Of course, Tesla isn't the only culprit in the car industry. And every company on earth appeals to our fantasies to sell their products. But for me, there's a difference between buying a perfume because I think it will make me more attractive, and buying a car because I think its full autonomy will keep me safe.

Marketing strategies aside, I can't help but wonder if we're thinking about driverless cars in the wrong way altogether.

By now, we know that humans are really good at understanding subtleties, at analysing context, applying experience and distinguishing patterns. We're really bad at paying attention, at precision, at consistency and at being fully aware of our surroundings. We have, in short, precisely the opposite set of skills to algorithms.

So, why not follow the lead of the tumour-finding software in the medical world and let the skills of the machine complement the skills of the human, and advance the abilities of both? Until we get to full autonomy, why not flip the equation on its head and aim for a self-driving system that supports the driver rather than the other way around? A safety net, like ABS or traction control, that can patiently monitor the road and stay alert for a danger the driver has missed. Not so much a chauffeur as a guardian.

That is the idea behind work being done by the Toyota research institute. They're building two modes into their car. There's the 'chauffeur' mode, which – like Audi's traffic-jam pilot – could take over in heavy congestion; and there's the 'guardian' mode, which runs in the background while a human drives, and acts as a safety net,[67] reducing the risk of an accident if anything crops up that the driver hasn't seen.

Volvo has adopted a similar approach. Its 'Autonomous Emergency Braking' system, which automatically slows the car down if it gets too close to a vehicle in front, is widely credited for the impressive safety record of the Volvo XC90. Since the car first went on sale in the UK in 2002, over 50,000 vehicles have been purchased, and not a single driver or passenger within any of them has been killed in a crash.[68]

Like much of the driverless technology that is so keenly discussed, we'll have to wait and see how this turns out. But one thing is for sure – as time goes on, autonomous driving will have a few lessons to teach us that apply well beyond the world of motoring. Not just about the messiness of handing over control, but about being realistic in our expectations of what algorithms can do.

If this is going to work, we'll have to adjust our way of thinking. We're going to need to throw away the idea that cars should work perfectly every time, and accept that, while mechanical failure might be a rare event, algorithmic failure almost certainly won't be any time soon.

So, knowing that errors are inevitable, knowing that if we proceed we have no choice but to embrace uncertainty, the conundrums within the world of driverless cars will force us to decide how good something needs to be before we're willing to let it loose on our streets. That's an important question, and it applies elsewhere. How good is good enough? Once you've built a flawed algorithm that *can* calculate something, should you let it?

Crime

IT WAS A WARM JULY day in 1995 when a 22-year-old university student packed up her books, left the Leeds library and headed back to her car. She'd spent the day putting the finishing touches to her dissertation and now she was free to enjoy the rest of her summer. But, as she sat in the front seat of her car getting ready to leave, she heard the sound of someone running through the multi-storey car park towards her. Before she had a chance to react, a man leaned in through the open window and held a knife to her throat. He forced her on to the back seat, tied her up, super-glued her eyelids together, took the wheel of the car and drove away.

After a terrifying drive, he pulled up at a grassy embankment. She heard a clunk as he dropped his seat down and then a shuffling as he started undoing his clothes. She knew he was intending to rape her. Fighting blind, she pulled her knees up to her chest and pushed outwards with all her might, forcing him backwards. As she kicked and struggled, the knife in his hand cut into his fingers and his blood dripped on to the seats. He hit her twice in the face, but then, to her immense relief, got out of the car and left. Two hours after her ordeal had begun, the student was found wandering down Globe Road in Leeds, distraught and dishevelled, her shirt torn, her face red from where he'd struck her and her eyelids sealed with glue.[1]

Sexual attacks on strangers like this one are incredibly rare, but when they do occur they tend to form part of a series. And sure enough, this wasn't the first time the same man had struck. When police analysed blood from the car they found its DNA matched a sample from a rape carried out in another multi-storey car park two years earlier. That attack had taken place some 100 kilometres further south, in Nottingham. And, after an appeal on the BBC *Crimewatch* programme, police also managed to link the case to three other incidents a decade before in Bradford, Leeds and Leicester.[2]

But tracking down this particular serial rapist was not going to be easy. Together, these crimes spanned an area of 7,046 square kilometres – an enormous stretch of the country. They also presented the police with a staggering number of potential suspects – 33,628 in total – each one of whom would have to be eliminated from their enquiries or investigated.[3]

An enormous search would have to be made, and not for the first time. The attacks ten years earlier had led to a massive man-hunt; but despite knocking on 14,153 front doors, and collecting numerous swabs, hair samples and all sorts of other evidence, the police investigation had eventually led nowhere. There was a serious risk that the latest search would follow the same path, until a Canadian ex-cop, Kim Rossmo, and his newly developed algorithm were brought in to help.[4]

Rossmo had a bold idea. Rather than taking into account the vast amount of evidence already collected, his algorithm would ignore virtually everything. Instead, it would focus its attention exclusively on a single factor: geography.

Perhaps, said Rossmo, a perpetrator doesn't randomly choose *where* they target their victims. Perhaps their choice of location isn't an entirely free or conscious decision. Even though these attacks had taken place up and down the country, Rossmo wondered if there could be an unintended pattern hiding in the geography of the crimes – a pattern simple enough to be exploited. There was a chance, he believed, that the locations at which crimes took place could betray where the criminal actually came from. The case of the serial rapist was a chance to put his theory to the test.

Operation Lynx and the lawn sprinkler

Rossmo wasn't the first person to suggest that criminals unwittingly create geographical patterns. His ideas have a lineage that dates back to the 1820s, when André-Michel Guerry, a lawyer-turned-statistician who worked for the French Ministry of Justice, started collecting records of the rapes, murders and robberies that occurred in the various regions of France.[5]

Although collecting these kinds of numbers seems a fairly standard thing to do now, at the time maths and statistics had only ever been applied to the hard sciences, where equations are used to elegantly describe the physical laws of the universe: tracing the path of a planet across a sky, calculating the forces within a steam engine – that sort of thing. No one had bothered to collect crime data before. No one had any idea what to count, how to count or how often they should count it. And anyway – people thought at the time – what was the point? Man was strong, independent in nature and wandering around acting according to his own free will. His behaviour couldn't possibly be captured by the paltry practice of statistics.[6]

But Guerry's analysis of his national census of criminals suggested otherwise. No matter where you were in France, he found, recognizable patterns appeared in what crimes were committed, how – and by whom. Young people committed more crimes than old, men more than women, poor more than rich. Intriguingly, it soon became clear that these patterns didn't change over time. Each region had its own set of crime statistics that would barely change year on year. With an almost terrifying exactitude, the numbers of robberies, rapes and murders would repeat themselves from one year to the next. And even the methods used by the murderers were predictable. This meant that Guerry and his colleagues could pick an area and tell you in advance exactly how many murders by knife, sword, stone, strangulation or drowning you could expect in a given year.[7]

So maybe it wasn't a question of the criminal's free will after all. Crime is not random; people are predictable. And it was precisely that predictability that, almost two centuries after Guerry's discovery, Kim Rossmo wanted to exploit.

Guerry's work focused on the patterns found at the country and regional levels, but even at the individual level, it turns out that people committing crime still create reliable geographical patterns. Just like the rest of us, criminals tend to stick to areas they are familiar with. They operate locally. That means that even the most serious of crimes will probably be carried out close to where the offender lives. And, as you move further and further away from the scene of the crime, the chance of finding your perpetrator's home slowly drops away,[8] an effect known to criminologists as 'distance decay'.

On the other hand, serial offenders are unlikely to target victims who live *very* close by, to avoid unnecessary police attention on their doorsteps or being recognized by neighbours. The result is known as a 'buffer zone' which encircles the offender's home, a region in which there'll be a very low chance of their committing a crime.[9]

These two key patterns – distance decay and the buffer zone – hidden among the geography of the most serious crimes, were at the heart of Rossmo's algorithm. Starting with a crime scene pinned on to a map, Rossmo realized he could mathematically balance these two factors and sketch out a picture of where the perpetrator might live.

That picture isn't especially helpful when only one crime has been committed. Without enough information to go on, the so-called *geoprofiling algorithm* won't tell you much more than good old-fashioned common sense. But, as more crimes are added, the picture starts to sharpen, slowly bringing into focus a map of the city that highlights areas in which you're most likely to catch your culprit.

It's as if the serial offender is a rotating lawn sprinkler. Just as it would be difficult to predict where the very next drop of water is going to fall, you can't foresee where your criminal will attack next. But once the water has been spraying for a while and many drops have fallen, it's relatively easy to observe from the pattern of the drops where the lawn sprinkler is likely to be situated.

And so it was with Rossmo's algorithm for Operation Lynx – the hunt for the serial rapist. The team now had the locations of five separate crimes, plus several places where a stolen credit card had been used by the attacker to buy alcohol, cigarettes and a video game. On the basis of just those locations, the algorithm highlighted two key areas in which it believed the perpetrator was likely to live: Millgarth and Killingbeck, both in the suburbs of Leeds.[10]

Back in the incident room, police had one other key piece of evidence to go on: a partial fingerprint left by the attacker at the scene of an earlier crime. It was too small a sample for an automatic fingerprint recognition system to be able to whizz through a database of convicted criminals' prints looking for a match, so any comparisons would need to be made meticulously by an expert with a magnifying glass, painstakingly examining one suspect at a time. By now the operation was almost three years old and – despite the best efforts of 180 different officers from five different forces – it was beginning to run out of steam. Every lead resulted in just another dead end.

Officers decided to manually check all the fingerprints recorded in the two places the algorithm had highlighted. First up was Millgarth: but a search through the prints stored in the local police database returned nothing. Then came Killingbeck – and after 940 hours of sifting through the records here, the police finally came up with a name: Clive Barwell.

Barwell was a 42-year-old married man and father of four, who had been in jail for armed robbery during the hiatus in the attacks. He now worked as a lorry driver and would regularly make long trips up and down the country in the course of his job; but he lived in Killingbeck and would often visit his mother in Millgarth, the two areas highlighted by the algorithm.[11] The partial print on its own hadn't been enough to identify him conclusively, but a subsequent DNA test proved that it was he who had committed these horrendous crimes. The police had their man. Barwell pleaded guilty in court in October 1999. The judge sentenced him to eight back-to-back life sentences.[12]

Once it was all over, Rossmo had the chance to take stock of how well the algorithm had performed. It had never actually pinpointed Barwell by name, but it did highlight on a map the areas where the police should focus their attention. If the police had used the algorithm to prioritize their list of suspects on the basis of where each of them lived – checking the fingerprints and taking DNA swabs of each one in turn – there would have been no need to trouble anywhere near as many innocent people. They would have found Clive Barwell after searching only 3 per cent of the area.[13]

This algorithm had certainly been proved effective. And it brought other positives, too. As it only prioritizes your existing list of suspects, it doesn't suffer from bias of the kind we met in the 'Justice' chapter. Also, it can't override good detective work, only make an investigation more efficient; so there's little chance of people putting too much trust in it.

It is also incredibly flexible. Since Operation Lynx, it has been used by more than 350 crime-fighting agencies around the world, including the US Federal Bureau of Investigation and the Royal Canadian Mounted Police. And the insights it offers extend beyond crime: the algorithm has been used to identify stagnant water pools that mosquitoes use as breeding grounds, based on the locations of cases of malaria in Egypt.[14] A PhD student of mine at University College London is currently using the algorithm in an attempt to predict the sites of bomb factories on the basis of the locations where improvised explosive devices are used. And one group of mathematicians in London have even used it to try to track down Banksy,[15] the reclusive street artist, on the basis of where his paintings have been found.

The kinds of crimes for which geoprofiling works best – serial rapes, murders and violent attacks – are, fortunately, rare. In reality the vast majority of infractions don't warrant the kind of man-hunt that the Clive Barwell case demanded. If algorithms were to make a difference in tackling crime beyond these extreme cases, they'd need a different geographical pattern to go on. One that could be applied to a city as a whole. One that could capture the patterns and rhythms of a street or a corner that every beat officer knows instinctively. Thankfully, Jack Maple had just the thing.

Charts of the Future

A lot of people would think twice about riding the New York City subway in the 1980s. It wasn't a nice place. Graffiti covered every surface, the cars stank of stale urine, and the platforms were rife with drug use, theft and robbery. Around 20 innocent people were murdered below ground every year, making it practically as dangerous a place as any in the world.

It was against this backdrop that Jack Maple worked as a police officer. He had recently earned himself a promotion to transit lieutenant, and in his years in the force he'd grown tired of only ever responding to crime, rather than fighting to reduce it. Out of that frustration was born a brilliant idea.

'On 55 feet of wall space, I mapped every train station in New York City and every train,' Maple told an interviewer in 1999. 'Then

I used crayons to mark every violent crime, robbery and grand larceny that occurred. I mapped the solved versus the unsolved.'[16]

They might not sound like much, but his maps, scrawled out on brown paper with crayons, became known as the 'Charts of the Future' and were, at the time, revolutionary. It had never occurred to anyone to look at crimes in that way before. But the moment Maple stepped back to view an entire city's worth of crime all at once, he realized he was seeing everything from a completely new perspective.

'It poses the question, Why?' he said. 'What are the underlying causes of why there is a certain cluster of crime in a particular place?' The problem was, at the time, every call to the police was being treated as an isolated incident. If you rang to report an aggressive group of drug dealers who were hanging out on a corner, but they ducked out of sight as soon as the cops arrived, nothing would be recorded that could link your complaint to other emergency calls made once the gang retook their positions. By contrast, Maple's maps meant he could pinpoint precisely where crime was a chronic problem, and that meant he could start to pick apart the causes. 'Is there a shopping centre here? Is that why we have a lot of pickpockets and robberies? Is there a school here? Is that why we have a problem at three o'clock? Is there an abandoned house nearby? Is that why there is crack-dealing on the corner?'[17]

Being able to answer these questions was the first step towards tackling the city's problems. So in 1990, when the open-minded Bill Bratton became head of the New York Transit Police, Maple showed him his Charts of the Future. Together they used them to try to make the subway a safer place for everyone.[18]

Bratton had a clever idea of his own. He knew that people begging, urinating and jumping the turnstiles were a big problem on the subway. He decided to focus police attention on addressing those minor misdemeanours rather than the much more serious crimes of robbery and murder that were also at epidemic levels below ground.

The logic was twofold. First, by being tough on any anti-social behaviour at the crime hotspots, you could send a strong signal that criminal activity was not acceptable in any form, and so hopefully start changing what people saw as 'normal'. Second, the people evading fares were disproportionately likely to be criminals who would then go on to commit bigger crimes. If they were arrested for fare evasion they wouldn't have that chance. 'By cracking down on fare evasion, we have been able to stop serious criminals carrying weapons at the turnstiles before they get on the subways and wreak havoc,' Bratton told *Newsday* in 1991.[19]

The strategy worked. As the policing got smarter, the subways got safer. Between 1990 and 1992, Maple's maps and Bratton's tactics cut felonies on the subway by 27 per cent and robberies by a third.[20]

When Bratton became Police Commissioner for the New York Police Department, he decided to bring Maple's Charts of the Future with him. There, they were developed and refined to become CompStat, a data-tracking tool that is now used by many police departments, both in the United States and abroad. At its core remains Jack Maple's simple principle – record where crimes have taken place to highlight where the worst hotspots are in a city.

Those hotspots tend to be quite focused. In Boston, for instance, a study that took place over a 28-year period found that 66 per cent of all street robberies happened on just 8 per cent of streets.[21] Another study mapped 300,000 emergency calls to police in Minneapolis: half of them came from just 3 per cent of the city area.[22]

But those hotspots don't stay the same over time. They constantly move around, subtly shifting and morphing like drops of oil on water, even from one day to the next. And when Bratton moved to Los Angeles in 2002, he began to wonder if there were other patterns that could tell you *when* crime was going to occur as well as *where*. Was there a way to look beyond crimes that had already happened? Rather than just responding to crime, the source of Maple's frustration, or fighting it as it happened, could you also predict it?

The flag and the boost

When it comes to predicting crime, burglary is a good place to start. Because burglaries take place at an address, you know precisely where they happened – unlike pickpocketing, say. After all, it's quite possible that a victim wouldn't notice their phone missing until they got home. Most people will report if they've been burgled, too, so we have really good, rich datasets, which are much harder to gather for drug-related offences, for instance. Plus, people often have a good idea of when their homes were burgled (perhaps when they were at work, or out for the evening), information you won't have for crimes such as vandalism.

Burglars also have something in common with the serial murderers and rapists that Rossmo studied: they tend to prefer sticking to areas they're familiar with. We now know you're more likely to be burgled if you live on a street that a burglar regularly uses, say on their way to work or school.[23] We* also know that there's a sweet spot for how busy burglars like a street to be: they tend to avoid roads with lots of traffic, yet home in on streets with lots of carefree non-locals walking around to act as cover (as long as there aren't also lots of nosey locals hanging about acting as guardians).[24]

But that's only the first of two components to the appeal of your home to a burglar. Yes, there are factors that don't change over time, like where you live or how busy your road is, that will 'flag' the steady

---

* On this occasion, when I say 'we' I actually mean it. This particular study was a paper I worked on with my marvellous PhD student Michael Frith.

appeal of your property as a target. But before you rush off to sell your house and move to a quiet cul-de-sac with a great Neighbourhood Watch scheme, you should also be aware that crime hotspots don't stay still. The second component of your home's appeal is arguably the more important. This factor depends on what exactly is going on right now in your immediate local neighbourhood. It's known as the 'boost'.

If you've ever been broken into twice within a short space of time, you'll be all too familiar with the boost effect. As police will tell you after you've first been victimized, criminals tend to repeatedly target the same location – and that means that no matter where you live, you're most at risk in the days right after you've just been burgled. In fact, your chances of being targeted can increase twelvefold at this time.[25]

There are a few reasons why burglars might decide to return to your house. Perhaps they've got to know its layout, or where you keep your valuables (things like TVs and computers are often replaced pretty quickly, too), or the locks on your doors, or the local escape routes; or it might be that they spotted a big-ticket item they couldn't carry the first time round. Whatever the reason, this boost effect doesn't just apply to you. Researchers have found that the chances of your neighbours being burgled immediately after you will also be boosted, as will those of your neighbours' neighbours, and your neighbours' neighbours' neighbours, and so on all the way down the street.

You can imagine these boosts springing up and igniting across the city like a fireworks display. As you get further away from the original spark, the boost gets fainter and fainter; and it fades away over time, too, until after two months – unless a new crime re-boosts the same street – it will have disappeared entirely.[26]

The flag and the boost in crime actually have a cousin in a natural phenomenon: earthquakes. True, you can't precisely predict where and when the first quake is going to hit (although you know that some places are more prone to them than others). But as soon as the first tremors start, you can talk quite sensibly about where and how often you expect the aftershocks to occur, with a risk that is highest at the site of the original quake, lessens as you get further away, and fades as time passes.

It was under Bratton's direction that the connection between earthquake patterns and burglary was first made. Keen to find a way to forecast crimes, the Los Angeles Police Department set up a partnership with a group of mathematicians at the University of California, Los Angeles, and let them dig through all the data the cops could lay their hands on: 13 million crime incidents over 80 years. Although criminologists had known about the flag and the boost for a few years by this point, in searching through the patterns in the data the UCLA group became the first to realize that the mathematical equations which so beautifully predicted the risk of seismic shocks and aftershocks could also be used to predict crimes and 'aftercrimes'. And it didn't just work for burglary. Here was a way to forecast everything from car theft to violence and vandalism.

The implications were indeed seismic. Rather than being able to say that a recently victimized area of the city was vaguely 'more at risk', with these equations, you could quantify exactly what that risk was, down to the level of a single street. And knowing, probabilistically speaking, that a particular area of the city would be the focus of burglars on a given night, it was easy to write an algorithm that could tell police where to target their attention.

And so PredPol (or PREDictive POLicing) was born.

The crime tipster

You might well have come across PredPol already. It's been the subject of thousands of news articles since its launch in 2011, usually under a headline referencing the Tom Cruise film *Minority Report*. It's become like the Kim Kardashian of algorithms: extremely famous, heavily criticized in the media, but without anyone really understanding what it does.

So before you fill your mind with images of seers lying in pools of water screaming out their premonitions, let me just manage your expectations slightly. PredPol doesn't track down people before they commit crimes. It can't target individual people at all, only geography. And I know I've been throwing the word 'prediction' around, but the algorithm can't actually tell the future. It's not a crystal ball. It can only predict the risk of future events, not the events themselves – and that's a subtle but important difference.

Think of the algorithm as something like a bookie. If a big group of police officers are crowded around a map of the city, placing bets on where crime will happen that night, PredPol calculates the odds. It acts like a tipster, highlighting the streets and areas that are that evening's 'favourites' in the form of little red squares on a map.

The key question is whether following the tipster's favourite can pay off. To test the algorithm's prowess,[27] it was pitted against the very best expert human crime analysts, in two separate experiments; one in Kent in southern England, the other in the Southwest Division of Los Angeles. The test was a straight head-to-head challenge. All the algorithm and the expert had to do was place 20 squares, each representing an area of 150 square metres, on a map to indicate where they thought most crime would happen in the next 12 hours.

Before we get to the results, it's important to emphasize just how tricky this is. If you or I were given the same task, assuming we don't have extensive knowledge of the Kentish or Californian crime landscapes, we'd probably do no better than chucking the squares on to the map at random. They'd cover pathetically little of it, mind – just one-thousandth of the total area in the case of Kent[28] – and every

12 hours you'd have to clear your previous guesses and start all over again. With this random scattering, we could expect to successfully 'predict' less than one in a hundred crimes.[29]

The experts did a lot better than that. In LA, the analyst managed to correctly predict where 2.1 per cent of crime would happen,[30] and the UK experts did better still with an average of 5.4 per cent,[31] a score that was especially impressive when you consider that their map was roughly ten times the size of the LA one.

But the algorithm eclipsed everyone. In LA it correctly forecast more than double the number of crimes that the humans had managed to predict, and at one stage in the UK test, almost one in five crimes occurred within the red squares laid down by the mathematics.[32] PredPol isn't a crystal ball, but nothing in history has been able to see into the future of crime so successfully.

Putting predictions into practice

But there's a problem. While the algorithm is relatively great at *predicting* where crime will happen over the next 12 hours, the police themselves have a subtly different objective: *reducing* crime over the next 12 hours. Once the algorithm has given you its predictions, it's not entirely clear what should be done next.

There are a few options, of course. In the case of burglary, you could set up CCTV cameras or undercover police officers and catch your criminals in the act. But perhaps it would be better for everyone if your efforts went into preventing the crime before it happened. After all, what would you prefer? Being the victim of a crime where the perpetrator gets caught? Or not being the victim of a crime in the first place?

You could warn local residents that their properties are at risk, maybe offer to improve the locks on their doors, maybe install burglar alarms or timers on their light switches to trick any dodgy people passing by into thinking there's someone at home. That's what one study did in Manchester in 2012,[33] where they managed to reduce the number of burglaries by more than a quarter. Small downside, though: the researchers calculated that this tactic of so-called 'target hardening' costs about £3,925 per burglary it prevents.[34] Try selling that to the Los Angeles Police Department, which deals with over 15,000 burglaries a year.[35]

Another option, which deviates as little as possible from traditional policing, is a tactic known as 'cops on the dots'.

'In the olden days,' Steyve Colgan, a retired Metropolitan Police officer, told me, '[patrols] were just geographic, you got the map, cut it up into chunks, and divvyed it up. You're on that beat, you're on that beat. As simple as that.' Problem was, as one UK study calculated, a police officer patrolling their randomly assigned beat on foot could expect to come within a hundred yards of a burglary just once in every eight years.[36]

With cops-on-the-dots, you simply send your patrols to the hotspots highlighted by the algorithm instead. (It should be called cops-on-the-hotspots, really. I guess that wasn't as catchy.) The idea being, of course, that if the police are as visible as possible and in the right place at the right time, they're more likely to stop crime from happening, or at least respond quickly in the immediate aftermath.

This is exactly what happened in Kent. During the second phase of the study, as the evening shift started, the sergeant printed off the maps with red squares highlighting the at-risk areas for that evening. Whenever the police on patrol had a quieter moment, they were to go to the nearest red box, get out of their cars and have a walk around.

On one particular evening, in an area they would never normally go to, police found an east European woman and her child in the street. It turned out that the woman was in an abusive relationship and, just minutes before, the child had been sexually assaulted. The sergeant on duty that night confirmed that 'they found these people because they were in a PredPol box'.[37] The suspect was arrested nearby later that night.

That mother and her child weren't the only people helped by the algorithm during the cops-on-the-dots trial: crime in Kent overall went down by 4 per cent. Similar studies in the United States (conducted by PredPol itself) report even bigger drops in crime. In the Foothill area of Los Angeles, crime dropped by 13 per cent in the first four months of using the algorithm, despite an increase of 0.4 per cent in the rest of the city, where they were relying on more traditional policing methods. And Alhambra, a city in California not far from LA, reported an extraordinary 32 per cent drop in burglaries and a 20 per cent drop in vehicle theft after deploying the algorithm in January 2013.[38]

These numbers are impressive, but it's actually difficult to know for sure whether PredPol can take the credit. Toby Davies, a mathematician and crime scientist from UCL, told me: 'It's possible that merely encouraging policing officers to go to places and get out of their cars and walk around, regardless of where, actually could lead to reductions [in crime] anyway.'

And there's another issue here. If, the harder you look for crime, the more likely you are to find it, then the act of sending police out could actually change the crime records themselves: 'When police are in a place,' Davies told me, 'they detect more crime than they would have done otherwise. Even if an equal value of crime is happening in two places, the police will detect more in the place they were than the one that they weren't.'

That means there is one very big potential downside of using a cops-on-the-dots tactic. By sending police into an area to fight crime on the back of the algorithm's predictions, you can risk getting into a feedback loop.

If, say, a poorer neighbourhood had a high level of crime in the first instance, the algorithm may well predict that more crime will happen there in future. As a result, officers are sent to the neigh-bourhood, which means they detect more crime. Thus, the algorithm predicts more still, more officers are sent there, and so on it goes. These feedback loops are more likely to be a problem for crimes that are linked to poorer areas such as begging, vagrancy and low-level drug use.

In the UK, where some sections of society regularly complain about a lack of police presence on the streets, focusing police attention on certain areas might not immediately seem unfair. But not everyone has a positive relationship with the police. 'It is legitimate for people who see a police officer walking in front of their house every day to feel oppressed by that, even if no one's doing any crimes, even if that police officer is literally just walking up and down,' Davies told me. 'You almost have a right not to be constantly under pressure, under the eye of the police.'

I'm rather inclined to agree.

Now, a well-tuned algorithm *should* be built so that it can take account of the tactics being used by the police. There are ways, theoretically at least, to ensure that the algorithm doesn't disproportionately target particular neighbourhoods – like randomly sending police to medium-risk areas as well as high-risk ones. But, unfortunately, there's no way to know for sure whether PredPol is managing to avoid these feedback loops entirely, or indeed whether it is operating fairly more generally, because PredPol is a proprietary algorithm, so the code isn't available to the public and no one knows exactly how it works.

PredPol is not the only software on the market. One competitor is HunchLab, which works by combining all sorts of statistics about an area: reported crimes, emergency calls, census data (as well as more eyebrow-raising metrics like moon phases). HunchLab doesn't have an underlying theory. It doesn't attempt to establish *why* crime occurs in some areas more than others; it simply reports on patterns it finds in the data. As a result, it can reliably predict more types of crime than PredPol (which has at its heart theories about how criminals create geographical patterns) – but, because HunchLab too is protected as intellectual property, it is virtually impossible from the outside to ensure it isn't inadvertently discriminating against certain groups of people.[39]

Another opaque predictive algorithm is the Strategic Subject List used by the Chicago Police Department.[40] This algorithm takes an entirely different approach from the others. Rather than focusing on geography, it tries to predict which individuals will be involved in gun crime. Using a variety of factors, it creates a 'heat list' of people it deems most likely to be involved in gun violence in the near future, either doing the shooting or being shot. The theory is sound: today's victims are often tomorrow's perpetrators. And the programme is well intentioned: officers visit people on the watch list to offer access to intervention programmes and help to turn their lives around.

But there are concerns that the Strategic Subject List might not be living up to its promise. One recent investigation by the non-profit RAND Corporation concluded that appearing on it actually made no difference to an individual's likelihood of being involved in a shooting.[41] It did, however, mean they were more likely to be arrested. Perhaps – the report concluded – this was because officers were simply treating the watch list as a list of suspects whenever a shooting occurred.

Predictive policing algorithms undoubtedly show promise, and the people responsible for creating them are undoubtedly doing so in good faith, with good intentions. But the concerns raised around bias and discrimination are legitimate. And for me, these questions are too fundamental to a just society for us simply to accept assurances that law enforcement agencies will use them in a fair way. It's one of many examples of how badly we need independent experts and a regulatory body to ensure that the good an algorithm does outweighs the harm.

And the potential harms go beyond prediction. As we have already seen in a variety of other examples, there is a real danger that algorithms can add an air of authority to an incorrect result. And the consequences here can be dramatic. Just because the computer says something doesn't make it so.

Who do you think you are?

Steve Talley was asleep at home in South Denver in 2014 when he heard a knock at the door.[42] He opened it to find a man apologizing for accidentally hitting his car. The stranger asked Talley to step outside and take a look. He obliged. As he crouched down to assess the damage to his driver's side door,[43] a flash grenade went off. Three men dressed in black jackets and helmets appeared and knocked him to the ground. One man stood on his face. Another restrained his arms while another started repeatedly hitting him with the butt of a gun.

Talley's injuries would be extensive. By the end of the evening he had sustained nerve damage, blood clots and a broken penis.[44] 'I didn't even know you could break a penis,' he later told a journalist at *The Intercept*. 'At one point I was actually screaming for the police. Then I realized these were cops who were beating me up.'[45]

Steve Talley was being arrested for two local bank robberies. During the second robbery a police officer had been assaulted, which is why, Talley thinks, he was treated so brutally during his arrest. 'I told them they were crazy,' he remembers shouting at the officers, 'You've got the wrong guy!'

Talley wasn't lying. His arrest was the result of his striking resemblance to the right guy – the real robber.

Although it was a maintenance man working in Talley's building who initially tipped off the police after seeing photos on the

Comparison chart 12

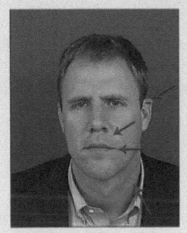

| K1 Images of Steven Talley | Q1 USA Bank Lobby camera image |

local news, it would eventually be an FBI expert using facial recognition software[46] who later examined the CCTV footage and concluded that 'the questioned individual depicted appears to be Talley'.[47]

Talley had a cast-iron alibi, but thanks to the FBI expert's testimony, it would still take over a year to clear his name entirely. In that time he was held in a maximum-security pod for almost two months until enough evidence surfaced to release him. As a result, he was unable to work, and by the time his ordeal was over he had lost his job, his home and access to his children. All as a direct result of that false identification.

Seeing double

Facial recognition algorithms are becoming commonplace in modern policing. These algorithms, presented with a photograph, footage or snapshot from a 3D camera, will detect a face, measure its characteristics and compare them to a database of known faces with the aim of determining the identity of the person pictured.

In Berlin, facial recognition algorithms capable of identifying known terrorism suspects are trained on the crowds that pass through railway stations.[48] In the United States, these algorithms have led to more than four thousand arrests since 2010 *just* for fraud and identity theft in the state of New York alone.[49] And in the UK, cameras mounted on vehicles that look like souped-up Google StreetView cars now drive around automatically cross-checking our likenesses with a database of wanted people.[50] These vans scored their first success in June 2017 after one drove past a man in south Wales where police had a warrant out for his arrest.[51]

Our safety and security often depend on our ability to identify and recognize faces. But leaving that task in the hands of humans can be risky. Take passport officers, for instance. In one recent study, set to mimic an airport security environment, these professional face recognizers failed to spot a person carrying the wrong ID a staggering 14 per cent of the time – and incorrectly rejected 6 per cent of perfectly valid matches.[52] I don't know about you, but I find those figures more than a little disconcerting when you consider the number of people passing through Heathrow every day.

As we shall see, facial recognition algorithms can certainly do better at the task than humans. But as they're applied to hunt for criminals, where the consequences of misidentification are so serious, their use raises an important question. Just how easily could one person's identity be confused with another's? How many of us have a Steve Talley-style lookalike lurking out there somewhere?

One study from 2015 seems to suggest the chances of you having your own real-life doppelgänger (whether they're a bank robber or otherwise) are vanishingly small. Teghan Lucas at the University of Adelaide painstakingly took eight facial measurements from photographs of four thousand people and failed to find a single match among them, leading her to conclude that the chances of two people having exactly the same face were less than one in a trillion.[53] By that calculation, Talley wasn't just 'a bit' unlucky. Taking into account that his particular one-in-a-trillion evil twin also lived nearby *and* happened to be a criminal, we could expect it to be tens of thousands of years before another ill-fated soul fell foul of the same miserable experience.

And yet there are reasons to suspect that those numbers don't quite add up. While it's certainly difficult to imagine meeting someone with the same face as yourself, anecdotal evidence of unrelated twin strangers does appear to be much more common than Lucas's research might suggest.

Take Neil Douglas, who was boarding a plane to Ireland when he realized his double was sitting in his seat. The selfie they took, with a plane-load of passengers laughing along in the background, quickly went viral, and soon redheads with beards from across the world were sending in photos of their own to demonstrate that they

too shared the likeness. 'I think there was a small army of us at some point,' Neil told the BBC.[54]

I even have my own story to add to the pile. When I was 22, a friend showed me a photo they'd seen on a local band's Myspace page. It was a collage of pictures taken at a gig that I hadn't attended, showing a number of people all enjoying themselves, one of whom looked eerily familiar. Just to be sure I hadn't unwittingly blacked out one night and wandered off to a party I now had no recollection of attending, I emailed the lead singer in the band, who confirmed what I suspected: my synth-pop-loving doppelgänger had a better social life than me.

So that's Talley, Douglas and me who each have at least one doppelgänger of our own, possibly more. We're up to three in a population of 7.5 billion and we haven't even started counting in earnest – and we're already way above Lucas's estimate of one in a trillion.

There is a reason for the discrepancy. It all comes down to the researcher's definition of 'identical'. Lucas's study required that two people's measurements must match one another exactly. Even though Neil and his lookalike are incredibly similar, if one nostril or one earlobe were out by so much as a millimetre, they wouldn't strictly count as doppelgängers according to her criteria.

But even when you're comparing two images of the same person, exact measurements won't reflect how each one of us is continually changing, through ageing, illness, tiredness, the expressions we're pulling or how our faces are distorted by a camera angle. Try to capture the essence of a face in millimetres and you'll find as much variation in one person's face as you will between people. Put simply, measurements alone can't distinguish one face from another.

Although they might not be perfectly identical, I can none the less easily imagine mixing up Neil and his twin-stranger in the photograph. Likewise in the Talley case – poor Steve didn't even look *that* similar to the real robber, and yet the images were misinterpreted by FBI experts to the point where he was charged with a crime he didn't commit and thrown into a maximum-security cell.

As the passport officers demonstrated, it's astonishingly easy to confuse unfamiliar faces, even when they bear only a passing resemblance. It turns out that humans are astonishingly bad at recognizing strangers. It's the reason why a friend of mine claimed she could barely sit through Christopher Nolan's beautifully made film *Dunkirk* – because she struggled to distinguish between the actors. It's why teenagers find it worthwhile to 'borrow' an older friend's ID to buy alcohol. And it's why the Innocence Project, a non-profit legal organization in the United States, estimates that eyewitness misidentification plays a role in more than 70 per cent of wrongful convictions.[55]

And yet, while an eyewitness might easily confuse Neil with his travel companion, his mother would surely have no problem picking out her son in the photo. When it comes to people we know, we are tremendously good at recognizing faces – even when it comes to real-life doppelgängers: a set of identical twins might be easily confused if they are only your acquaintances, but just as easily distinguished once you know them properly.

And herein lies a critical point: similarity is in the eye of the beholder. With no strict definition of similarity, you can't measure *how* different two faces are and there is no threshold at which we can say that two faces are identical. You can't define what it means to be a doppelgänger, or say how common a particular face is; nor – crucially – can you state a probability that two images were taken from the same individual.

This means that facial recognition, as a method of identification, is not like DNA, which sits proudly on a robust statistical platform. When DNA testing is used in forensics, the profiling focuses on particular chunks of the genome that are known to be highly variable between humans. The extent of that variation is key: if the DNA sequence in a sample of body tissue found at the scene of a crime matches the sequence in a swab from a suspect, it means you can calculate the probability that both came from the same individual. It also means you can state the exact chance that some unlucky soul just happened to have an identical DNA sequence at those points.[56] The more markers you use, the lower your chances of a mismatch, and so, by choosing the number of markers to test, every judicial system in the world has complete power to decide on the threshold of doubt they're willing to tolerate.[57]

Even though our faces feel so intrinsically linked to who we are, without knowing the variation across humans the practice of identifying felons by their faces isn't supported by rigorous science. When it comes to identifying people from photos – to quote a presentation given by an FBI forensics unit – 'Lack of statistics means: conclusions are ultimately opinion based.'[58]

Unfortunately, using algorithms to do our facial recognition for us does not solve this conundrum, which is one very good reason to exercise caution when using them to pinpoint criminals. Resemblance and identity are not the same thing and never will be, however accurate the algorithms become.

And there's another good reason to tread carefully with face-recognition algorithms. They're not quite as good at recognizing faces as you might think.

One in a million?

The algorithms themselves work using one of two main approaches. The first kind builds a 3D model of your face, either by combining a series of 2D images or by scanning you using a special infrared camera. This is the method adopted by the Face ID system that Apple uses in its iPhones. These algorithms have worked out a way to get around the issues of different facial expressions and ageing by focusing on areas of the face that have rigid tissue and bone, like the curve of your eye socket or the ridge of your nose.

Apple has claimed that the chance of a random person being able to unlock your phone with Face ID is one in a million, but the algorithm is not flawless. It can be fooled by twins,[59] siblings,[60] and children on their parents' phones. (Soon after the launch of Face ID, a video appeared of a ten-year-old boy who could hoodwink the facial recognition on his mother's iPhone. She now deletes her texts if there is something she doesn't want her son to look at.)[61] There have also been reports that the algorithm can be tricked by a specially built 3D printed mask, with infrared images glued on for the eyes.[62] All this means that while the algorithm might be good enough to unlock your phone, it probably isn't yet reliable enough to be used to grant access to your bank accounts.

Nor are these 3D algorithms much use for scanning passport photos or CCTV footage. For that you need the second kind of algorithm, which sticks to 2D images and uses a statistical approach. These algorithms don't directly concern themselves with landmarks that you or I could recognize as distinguishing features, but instead build a statistical description of the patterns of light and dark across the image. Like the algorithms built to recognize dogs in the 'Medicine' chapter, researchers realized recently that, rather than having to rely on humans to decide which patterns will work best, you can get the algorithm to learn the best combinations for itself, by using trial and error on a vast dataset of faces. Typically, it's done using neural networks. This kind of algorithm is where the big recent leaps forward in performance and accuracy have come in. That performance, though, comes with a cost. It isn't always clear precisely *how* the algorithm decides whether one face is like another.

That means these state-of-the-art algorithms can also be pretty easily fooled. Since they work by detecting a statistical description

of the patterns of light and dark on a face, you can trick them just by wearing funky glasses with a disruptive pattern printed on them. Even better, by designing the specific disruptive pattern to signal someone else's face, you can actually make the algorithm think you are that person – as the chap in the image above is doing, wearing glasses that make him look 'like' actress Milla Jovovich.[63] Using glasses as a disguise? Turns out Clark Kent was on to something.

But, targeted attacks with funky glasses aside, the recognition abilities of these statistical algorithms have prompted many admiring headlines, like those that greeted Google's FaceNet. To test its recognition skills, FaceNet was asked to identify five thousand images of celebrities' faces. Human recognizers had previously attempted the same task and done exceptionally well, scoring 97.5 per cent correct identifications (unsurprisingly, since these celebrity faces would have been familiar to the participants).[64] But FaceNet did even better, scoring a phenomenal 99.6 per cent correct.

On the surface, this looks as if the machines have mastered superhuman recognition skills. It sounds like a great result, arguably good enough to justify the algorithms being used to identify criminals. But there's a catch. Five thousand faces is, in fact, a pathetically small number to test your algorithm on. If it's going to be put to work fighting crime, it's going to need to find one face among millions, not thousands.

That's because the UK police now hold a database of 19 million images of our faces, created from all those photos taken of individuals arrested on suspicion of having committed a crime. The FBI, meanwhile, has a database of 411 million images, in which half of all American adults are reportedly pictured.[65] And in China, where the ID card database gives easy access to billions of faces, the authorities have already invested heavily in facial recognition. There are cameras installed in streets, subways and airports that will supposedly spot everything from wanted criminals to jaywalkers as they travel through the country's cities.[66] (There's even a suggestion that a citizen's minor misdemeanours in the physical world, like littering, will form part of their Sesame Credit score – attracting all of the associated punishments that we uncovered in the 'Data' chapter.)

Here's the problem: the chances of misidentification multiply dramatically with the number of faces in the pile. The more faces the algorithm searches through, the more chance it has of finding two faces that look similar. So, once you try using these same algorithms on bigger catalogues of faces, their accuracy plummets.

It would be a bit like getting me to match ID cards to ten strangers and – when I got full marks – claiming that I was capable of correctly identifying faces 100 per cent of the time, then letting me wander off into the centre of New York to identify known criminals. It's inevitable that my accuracy would drop.

It's just the same with the algorithms. In 2015, the University of Washington set up the so-called MegaFace challenge, in which people from around the world were invited to test their recognition algorithms on a database of 1 million faces.[67] Still substantially smaller than the catalogues held by some government authorities, but getting closer. Even so, the algorithms didn't handle the challenge well.

Google's FaceNet – which had been close to perfect on the celebrities – could suddenly manage only a 75 per cent* identification rate.[68] Other algorithms came in at a frankly pathetic 10 per cent success rate. At the time of writing, the world's best is a Chinese offering called Tencent YouTu Lab, which can manage an 83.29 per cent recognition rate.[69]

To put that another way, if you're searching for a particular criminal in a digital line-up of millions, based on those numbers, the *best-case scenario* is that you won't find the right person one in six times.

Now, I should add that progress in this area is happening quickly. Accuracy rates are increasing steadily, and no one can say for certain what will happen in the coming years or months. But I can tell you that differences in lighting, pose, image quality and general appearance make accurately and reliably recognizing faces a very tricky problem indeed. We're some way away from getting perfect accuracy on databases of 411 million faces, or being able to find that one-in-a-trillion doppelgänger match.

Striking a balance

These are sobering facts, but not necessarily deal-breakers. There are algorithms good enough to be used in some situations. In Ontario, Canada, for instance, people with a gambling addiction can voluntarily place themselves on a list that bars them from entering a casino. If their resolve wavers, their face will be flagged by

---

* This number relates to how many true matching faces the algorithm missed when it was tuned to avoid misidentification. There's more on the kinds of errors an algorithm can make in the 'Justice' chapter, and the different ways to measure accuracy in the 'Medicine' chapter.

recognition algorithms, prompting casino staff to politely ask them to leave.[70] The system is certainly unfair on all those mistakenly prevented from a fun night on the roulette table, but I'd argue that's a price worth paying if it means helping a recovering gambling addict resist the temptation of their old ways.

Likewise in retail. In-store security guards used to have offices plastered with Polaroids of shoplifters; now algorithms can cross-reference your face with a database of known thieves as soon as you pass the threshold of the store. If your face matches that of a well-known culprit, an alert is sent to the smartphones of the guards on duty, who can then hunt you out among the aisles.

There's good reason for stores to want to use this kind of technology. An estimated 3.6 million offences of retail crime are committed every year in the UK alone, costing retailers a staggering £660 million.[71] And, when you consider that in 2016 there were 91 violent deaths of shoplifting suspects at retail locations in the United States,[72] there is an argument that a method of preventing persistent offenders from entering a store before a situation escalates would be good for everyone.

But this high-tech solution to shoplifting comes with downsides: privacy, for one thing (FaceFirst, one of the leading suppliers of this kind of security software, claims it doesn't store the images of regular customers, but shops are certainly using facial recognition to track our spending habits). And then there's the question of who ends up on the digital blacklist. How do you know that everyone on the list is on there for the right reasons? What about innocent until proven guilty? What about people who end up on the list accidentally: how do they get themselves off it? Plus again there's the potential for mis-identification by an algorithm that can never be perfectly accurate.

The question is whether the pros outweigh the cons. There's no easy answer. Even retailers don't agree. Some are enthusiastically adopting the technology, while others are moving away from it – including Walmart, which cancelled a FaceFirst trial in their stores after it failed to offer the return on investment the company were hoping for.[73]

But in the case of crime the balance of harm and good feels a lot more clear cut. True, these algorithms aren't alone in their slightly shaky statistical foundations. Fingerprinting has no known error rate either,[74] nor do bite mark analysis, blood spatter patterning[75] or ballistics.[76] In fact, according to a 2009 paper by the US National Academy of Sciences, none of the techniques of forensic science apart from DNA testing can 'demonstrate a connection between evidence and a specific individual or source'.[77] None the less, no one can deny that they have all proved to be incredibly valuable police tools – just as long as the evidence they generate isn't relied on too heavily. But the accuracy rates of even the most sophisticated facial recognition algorithms leave a lot to be desired. There's an argument that if there is even a slight risk of more cases like Steve Talley, then a technology that isn't perfect shouldn't be used to assist in robbing someone of their freedom. The only problem is that stories like Talley's don't quite paint the entire picture. Because, while there are enormous downsides to using facial recognition to catch criminals, there are also gigantic upsides.

The tricky trade-off

In May 2015, a man ran through the streets of Manhattan randomly attacking passers-by with a black claw hammer. First, he ran up to a group of people near the Empire State Building and smashed a 20-year-old man in the back of the head. Six hours later he headed south to Union Square and, using the same hammer, attacked a woman quietly sitting on a park bench on the side of the head. Just a few minutes later he appeared again, this time targeting a 33-year-old woman walking down the street outside the park.[78] Using surveillance footage from the attacks, a facial recognition algorithm was able to identify him as David Baril, a man who, months before the attacks, had posted a picture on Instagram of a hammer dripping with blood.[79] He pleaded guilty to the charges stemming from the attacks and was sentenced to 22 years in prison.

Cold cases, too, are being re-ignited by facial recognition breakthroughs. In 2014, an algorithm brought to justice an American man who had been living as a fugitive under a fake name for 15 years. Neil Stammer had absconded while on bail for charges including child sex abuse and kidnapping; he was re-arrested when his FBI 'Wanted' poster was checked against a database of passports and found to match a person living in Nepal whose passport photo carried a different name.[80]

After the summer of 2017, when eight people died in a terrorist attack on London Bridge, I can appreciate how helpful a system that used such an algorithm might be. Youssef Zaghba was one of three men who drove a van into pedestrians before launching into a stabbing spree in neighbouring Borough Market. He was on a watch list for terrorist suspects in Italy, and could have been automatically identified by a facial recognition algorithm before he entered the country.

But how do you decide on that trade-off between privacy and protection, fairness and safety? How many Steve Talleys are we willing to accept in exchange for quickly identifying people like David Baril and Youssef Zaghba?

Take a look at the statistics provided by the NYPD. In 2015, it reported successfully identifying 1,700 suspects leading to 900 arrests, while mismatching five individuals.[81] Troubling as each and every one of those five is, the question remains: is that an acceptable ratio? Is that a price we're willing to pay to reduce crime?

As it turns out, algorithms without downsides, like Kim Rossmo's geoprofiling, discussed at the beginning of the chapter, are the exception rather than the rule. When it comes to fighting crime, every way you turn you'll find algorithms that show great promise in one regard, but can be deeply worrying in another. PredPol, HunchLab, Strategic Subject Lists and facial recognition – all promising to solve all our problems, all creating new ones along the way.

To my mind, the urgent need for algorithmic regulation is never louder or clearer than in the case of crime, where the very existence of these systems raises serious questions without easy answers. Somehow, we're going to have to confront these difficult dilemmas. Should we insist on only accepting algorithms that we can understand or look inside, knowing that taking them out of the hands of their proprietors might mean they're less effective (and crime rates rise)? Do we dismiss any mathematical system with built-in biases, or proven capability of error, knowing that in doing so we'd be holding our algorithms to a higher standard than the human system we're left with? And how biased is too biased? At what point do you prioritize the victims of preventable crimes over the victims of the algorithm?

In part, this comes down to deciding, as a society, what we think success looks like. What is our priority? Is it keeping crime as low as possible? Or preserving the freedom of the innocent above all else? How much of one would you sacrifice for the sake of the other?

Gary Marx, professor of sociology at MIT, put the dilemma well in an interview he gave to the *Guardian*: 'The Soviet Union had remarkably little street crime when they were at their worst of their totalitarian, authoritarian controls. But, my God, at what price?'[82]

It may well be that, in the end, we decide that there should be some limits to the algorithm's reach. That some things should not be analysed and calculated. That might well be a sentiment that eventually applies beyond the world of crime. Not, perhaps, for lack of trying by the algorithms themselves. But because – just maybe – there are some things that lie beyond the scope of the dispassionate machine.

Art

JUSTIN WAS IN A REFLECTIVE mood. On 4 February 2018, in the living room of his home in Memphis, Tennessee, he sat watching the Super Bowl, eating M&Ms. Earlier that week he'd celebrated his 37th birthday, and now – as had become an annual tradition – he was brooding over what his life had become.

He knew he should be grateful, really. He had a perfectly comfortable life. A stable nine-to-five office job, a roof over his head and a family who loved him. But he'd always wanted something more. Growing up, he'd always believed he was destined for fame and fortune.

So how had he ended up being so ... normal? 'It was that boyband,' he thought to himself. The one he'd joined at 14. 'If we'd been a hit, everything would have been different.' But, for whatever reason, the band was a flop. Success had never quite happened for poor old Justin Timberlake.

Despondent, he opened another beer and imagined what might have been. On the screen, the Super Bowl commercials came to an end. Music started up for the big half-time show. And in a parallel universe – virtually identical to this one in all but one detail – another 37-year-old Justin Timberlake from Memphis took the stage.

Many worlds

Why is the real Justin Timberlake so successful? And why did the other Justin Timberlake fail? Some people (my 14-year-old-self included)* might argue that pop-star Justin's success is deserved: his natural talent, his good looks, his dancing abilities and the artistic merit of his music made fame inevitable. But others might disagree. Perhaps they'd claim there is nothing particularly special about Timberlake, or any of the other pop superstars who are worshipped by legions of fans. Finding talented people who can sing and dance is easy – the stars are just the ones who got lucky.

There's no way to know for sure, of course. Not without building a series of identical parallel worlds, releasing Timberlake into each and watching all the incarnations evolve, to see if he manages success every time. Unfortunately, creating an artificial multiverse is beyond most of us, but if you set your sights below Timberlake and consider less well-known musicians instead, it is still possible to explore the relative roles of luck and talent in the popularity of a hit record.

This was precisely the idea behind a famous experiment conducted by Matthew Salganik, Peter Dodds and Duncan Watts back in 2006 that created a series of digital worlds.[1] The scientists built their own online music player, like a very crude version of Spotify, and filtered visitors off into a series of eight parallel musical websites, each identically seeded with the same 48 songs by undiscovered artists.

In what became known as the Music Lab,[2] a total of 14,341 music fans were invited to log on to the player, listen to clips of each track, rate the songs, and download the music they liked best.

Just as on the real Spotify, visitors could see at a glance what music other people in their 'world' were listening to. Alongside the artist name and song title, participants saw a running total of how many times the track had already been downloaded within their world.

---

* Then again, my 14-year-old-self was also a big fan of PJ and Duncan, so what does she know?

All the counters started off at zero, and over time, as the numbers changed, the most popular songs in each of the eight parallel charts gradually became clear.

Meanwhile, to get some natural measure of the 'true' popularity of the records, the team also built a control world, where visitors' choices couldn't be influenced by others. There, the songs would appear in a random order on the page – either in a grid or in a list – but the download statistics were shielded from view.

The results were intriguing. All the worlds agreed that some songs were clear duds. Other songs were stand-out winners: they ended up being popular in every world, even the one where visitors couldn't see the number of downloads. But in between sure-fire hits and absolute bombs, the artists could experience pretty much any level of success.

Take 52Metro, a Milwaukee punk band, whose song 'Lockdown' was wildly popular in one world, where it finished up at the very top of the chart, and yet completely bombed in another world, ranked 40th out of 48 tracks. Exactly the same song, up against exactly the same list of other songs; it was just that in this particular world, 52Metro never caught on.[3] Success, sometimes, was a matter of luck.

Although the path to the top wasn't set in stone, the researchers found that visitors were much more likely to download tracks they knew were liked by others. If a middling song got to the top of the charts early on by chance, its popularity could snowball. More downloads led to more downloads. Perceived popularity became real popularity, so that eventual success was just randomness magnified over time.

There was a reason for these results. It's a phenomenon known to psychologists as *social proof*. Whenever we haven't got enough information to make decisions for ourselves, we have a habit of copying the behaviour of those around us. It's why theatres sometimes secretly plant people in the audience to clap and cheer at the right times. As soon as we hear others clapping, we're more likely to join in. When it comes to choosing music, it's not that we necessarily have a preference for listening to the same songs as others, but that popularity is a quick way to insure yourself against disappointment. 'People are faced with too many options,' Salganik told *LiveScience* at the time. 'Since you can't listen to all of them, a natural short cut is to listen to what other people are listening to.'[4]

We use popularity as a proxy for quality in all forms of entertainment. For instance, a 2007 study looked into the impact of an appearance in the *New York Times* bestseller list on the public perception of a book. By exploiting the idiosyncrasies in the way the list is compiled, Alan Sorensen, the author of the study, tracked the success of books that should have been included on the basis of their actual sales, but – because of time lags and accidental omissions – weren't, and compared them to those that did make it on to the list. He found a dramatic effect: just being on the list led to an increase in sales of 13–14 per cent on average, and a 57 per cent increase in sales for first-time authors.

The more platforms we use to see what's popular – bestseller lists, Amazon rankings, Rotten Tomatoes scores, Spotify charts – the bigger the impact that social proof will have. The effect is amplified further when there are millions of options being hurled at us, plus marketing, celebrity, media hype and critical acclaim all demanding your attention.

All this means that sometimes terrible music can make it to the top. That's not just me being cynical. During the 1990s, two British music producers – fully aware of this fact – were rumoured to have made a bet on who could get the worst song possible into the charts. Supposedly, the result of the wager was a girl group called Vanilla, whose debut song, 'No way no way, mah na mah na', was based on the famous Muppets ditty. It featured a performance that could only generously be described as singing, artwork that looked like it had been made in Microsoft Paint and a promotional video that had a good claim to being the worst ever shown.[5] But Vanilla had some powerful allies. Thanks to a few magazine features and an appearance on BBC's *Top of the Pops*, the song still managed to get to number 14 in the charts.*

Admittedly, the band's success was short-lived. By their second single, their popularity was already waning. They never released a third. All of which does seem to suggest that social proof isn't the only factor at play – as indeed a follow-up experiment from the Music Lab team showed.

The set-up to their second study was largely the same as the first. But this time, to test how far the perception of popularity became a self-fulfilling prophecy, the researchers added a twist. Once the charts had had the chance to stabilize in each world, they paused the experiment and flipped the billboard upside down. New visitors to the music player saw the chart topper listed at the bottom, while the flops at the bottom took on the appearance of the crème de la crème at the top.

Almost immediately, the total number of downloads by visitors dropped. Once the songs at the top weren't appealing, people lost interest in the music on the website overall. The sharpest declines were in downloads for the turkeys, now at the top of the charts. Meanwhile, the good tracks languishing at the bottom did worse than when they were at the top, but still better than those that had previously been at the end of the list. If the scientists had let the experiment run on long enough, the very best songs would have recovered their popularity. Conclusion: the market isn't locked into a particular state. Both luck and quality have a role to play.[6]

---

* It's unfortunate that this book's format doesn't lend itself to include musical excerpts, because I *really* want you to hear how hilariously bad this song is. Google it, will you?

Back in reality – where there is only one world's worth of data to go on – there's a straightforward interpretation of the findings from the Music Lab experiments. Quality matters, and it's not the same thing as popularity. That the very best songs recovered their popularity shows that some music is just inherently 'better'. At one end of the spectrum, a sensational song by a fantastic artist should (at least in theory) be destined for success. But the catch is that the reverse doesn't necessarily hold. Just because something is successful, that doesn't mean it's of a high quality.

Quite how you define quality is another matter altogether, which we'll come on to in a moment. But for some, quality itself isn't necessarily important. If you're a record label, or a film producer, or a publishing house, the million-dollar question is: can you spot the guaranteed successes in advance? Can an algorithm pick out the hits?

Hunting the hits

Investing in movies is a risky business. Few films make money, most will barely break even, and flops are part of the territory.[7] It's a high-stakes business: when the costs of making a movie run into the tens or hundreds of millions, failure to predict the demand for the product can be catastrophically expensive.

That was a lesson learned the hard way by Disney with its film *John Carter*, released in 2012. The studio sank $350 million into making the movie, determined that it should sit alongside the likes of *Toy Story* and *Finding Nemo* as their next big franchise. Haven't seen it? Me neither. The film failed to capture the public's imagination and wound up making a loss of $200 million, resulting in the resignation of the head of Walt Disney Studios.[8]

The great and the good of Hollywood have always accepted that you just can't accurately predict the commercial success of a movie.

It's the land of the gut-feel. Gambling on films that might bomb in the box office is just part of the job. In 1978, Jack Valenti, president and CEO of the Motion Picture Association of America, put it this way: 'No one can tell you how a movie is going to do in the market-place. Not until the film opens in a darkened theatre and sparks fly up between the screen and the audience.'[9] Five years later, in 1983, William Goldman – the writer behind *The Princess Bride* and *Butch Cassidy and the Sundance Kid* – put it more succinctly: 'Nobody knows anything.'[10]

But, as we've seen throughout this book, modern algorithms are routinely capable of predicting the seemingly unpredictable. Why should films be any different? You can measure the success of a movie, in revenue and in critical reception. You can measure all sorts of factors about the structure and features of a film: starring cast, genre, budget, running time, plot features and so on. So why not apply these same techniques to try and find the gems? To un-cover which films are destined to triumph in the box office?

This has been the ambition of a number of recent scientific stud-ies that aim to tap into the vast, rich depths of information collected and curated by websites like the Internet Movie Database (IMDb) or Rotten Tomatoes. And – perhaps unsurprisingly – there are a number of intriguing insights hidden within the data.

Take the study conducted by Sameet Sreenivasan in 2013.[11] He realized that, by asking users to tag films with plot keywords, IMDb had created a staggeringly detailed catalogue of descriptors that could show how our taste in films has evolved over time. By the time of his study, IMDb had over 2 million films in its catalogue, spanning more than a century, each with multiple plot tags. Some keywords were high-level descriptions of the movie, like 'organized-crime' or 'father-son-relationship'; others would be location-based, like 'manhattan-new-york-city', or about specific plot points, like 'held-at-gunpoint' or 'tied-to-a-chair'.

On their own, the keywords showed that our interest in certain plot elements tends to come in bursts; think Second World War films or movies that tackle the subject of abortion. There'll be a spate of releases on a similar topic in quick succession, and then a lull for a while. When considered together, the tags allowed Sreenivasan to come up with a score for the novelty of each film at the time of its release – a number between zero and one – that could be compared against box-office success.

If a particular plot point or key feature – like female nudity or organized crime – was a familiar aspect of earlier films, the keyword would earn the movie a low novelty score. But any original plot characteristics – like the introduction of martial arts in action films in the 1970s, say – would earn a high novelty score when the characteristic first appeared on the screen.

As it turns out, we have a complicated relationship with novelty. On average, the higher the novelty score a film had, the better it did at the box office. But only up to a point. Push past that novelty threshold and there's a precipice waiting; the revenue earned by a film fell off a cliff for anything that scored over 0.8. Sreenivasan's study showed what social scientists had long suspected: we're put off by the banal, but also hate the radically unfamiliar. The very best films sit in a narrow sweet spot between 'new' and 'not too new'.

The novelty score might be a useful way to help studios avoid backing absolute stinkers, but it's not much help if you want to know the fate of an individual film. For that, the work of a European team of researchers may be more useful. They discovered a connection between the number of edits made to a film's Wikipedia page in the month leading up to its cinematic release and the eventual box-office takings.[12] The edits were often made by people unconnected to the release – just typical movie fans contributing information to the page. More edits implied more buzz around a release, which in turn led to higher takings at the box office.

Their model had modest predictive power overall: out of 312 films in the study, they correctly forecast the revenue of 70 movies with an accuracy of 70 per cent or over. But the better a film did, and the more edits were made to the Wikipedia page, the more data the team had to go on and the more precise the predictions they made. The box-office takings of six high-earning films were correctly forecast to 99 per cent accuracy.

These studies are intellectually interesting, but a model that works only a month before a film's release isn't much use for investors. How about tackling the question head-on instead: take all the factors that are known earlier in the process – the genre, the celebrity status of the leading actors, the age guidance rating (PG, 12, etc.) – and use a machine-learning algorithm to predict whether a film will be a hit?

One famous study from 2005 did just that, using a neural network to try to predict the performance of films long before their release in the cinema.[13] To make things as simple as possible, the authors did away with trying to forecast the revenue exactly, and instead tried to classify movies into one of nine categories, ranging from total flop to box-office smash hit. Unfortunately, even with that step to simplify the problem, the results left a lot to be desired. The neural network outperformed any statistical techniques that had been tried before, but still managed to classify the performance of a movie correctly only 36.9 per cent of the time on average. It was a little better in the top category – those earning over $200 million – correctly identifying those real blockbusters 47.3 per cent of the time. But investors beware. Around 10 per cent of the films picked out by the algorithm as destined to be hits went on to earn less than $20 million – which by Hollywood's standards is a pitiful amount.

Other studies since have tried to improve on these predictions, but none has yet made a significant leap forward. All the evidence points in a single direction; until you have data on the early audience reaction, popularity is largely unpredictable. When it comes to picking the hits from the pile, Goldman was right. Nobody knows anything.

Quantifying quality

So predicting popularity is tricky. There's no easy way to prise apart what we all like from why we like it. And that poses rather a problem for algorithms in the creative realm. Because if you can't use popularity to tell you what's 'good' then how can you measure quality?

This is important: if we want algorithms to have any kind of autonomy within the arts – either to create new works, or to give us meaningful insights into the art we create ourselves – we're going to need some kind of measure of quality to go on. There has to be an objective way to point the algorithm in the right direction, a 'ground truth' that it can refer back to. Like an art analogy of 'this cluster of cells is cancerous' or 'the defendant went on to commit a crime'. Without it, making progress is tricky. We can't design an algorithm to compose or find a 'good' song if we can't define what we mean by 'good'.

Unfortunately, in trying to find an objective measure of quality, we come up against a deeply contentious philosophical question that dates back as far as Plato. One that has been the subject of debate for more than two millennia. How do you judge the aesthetic value of art?

Some philosophers – like Gottfried Leibniz – argue that if there are objects that we can all agree on as beautiful, say Michelangelo's *David* or Mozart's *Lacrimosa*, then there should be some definable, measurable, essence of beauty that makes one piece of art objectively better than another.

But on the other hand, it's rather rare for everyone to agree. Other philosophers, such as David Hume, argue that beauty is in the eye of the beholder. Consider the work of Andy Warhol, for instance, which offers a powerful aesthetic experience to some, while others find it artistically indistinguishable from a tin of soup.

Others still, Immanuel Kant among them, have said the truth is something in between. That our judgements of beauty are not wholly subjective, nor can they be entirely objective. They are sensory, emotional and intellectual all at once – and, crucially, can change over time depending on the state of mind of the observer.

There is certainly some evidence to support this idea. Fans of Banksy might remember how he set up a stall in Central Park, New York, in 2013, anonymously selling original black-and-white spray-painted canvases for $60 each. The stall was tucked away in a row of others selling the usual touristy stuff, so the price tag must have seemed expensive to those passing by. It was several hours before someone decided to buy one. In total, the day's takings were $420.[14] But a year later, in an auction house in London, another buyer would deem the aesthetic value of the very same artwork great enough to tempt them to spend £68,000 (around $115,000 at the time) on a single canvas.[15]

Admittedly, Banksy isn't popular with everyone. (Charlie Brooker – creator of *Black Mirror* – once described him as 'a guff-head [whose] work looks dazzlingly clever to idiots'.)[16] So you might argue this story is merely evidence of the fact that Banksy's work doesn't have inherent quality. It's just popular hype (and social proof) that drives those eye-wateringly high prices. But our fickle aesthetic judgement has also been observed in respect of art forms that are of undeniably high quality.

My favourite example comes from an experiment conducted by the *Washington Post* in 2007.[17] The paper asked the internationally renowned violinist Joshua Bell to add an extra concert to his schedule of sold-out symphony halls. Armed with his $3.5 million Stradivarius violin, Bell pitched up at the top of an escalator in a metro station in Washington DC during morning rush hour, put a hat on the ground to collect donations and performed for 43 minutes. As the *Washington Post* put it, here was one of 'the finest classical musicians in the world, playing some of the most elegant music ever written on one of the most valuable violins ever made'. The result? Seven people stopped to listen for a while. Over a thousand more walked straight past. By the end of his performance, Bell had collected a measly $32.17 in his hat.

What we consider 'good' also changes. The appetite for certain types of classical music has been remarkably resilient to the passing of time, but the same can't be said for other art forms. Armand Leroi, a professor of evolutionary biology at Imperial College London, has studied the evolution of pop music, and found clear evidence of our changing tastes in the analysis. 'There's an intrinsic boredom threshold in the population. There's just a tension that builds as people need something new.'[18]

By way of an example, consider the drum machines and synthesizers that became fashionable in late-1980s pop – so fashionable that the diversity of music in the charts plummeted. 'Everything sounds like early Madonna or something by Duran Duran,' Leroi explains. 'And so maybe you say, "OK. We've reached the pinnacle of pop. That's where it is. The ultimate format has been found."' Except, of course, it hadn't. Shortly afterwards, the musical diversity of the charts exploded again with the arrival of hip hop. Was there something special about hip hop that caused the change? I asked Leroi. 'I don't think so. It could have been something else, but it just happened to be hip hop. To which the American consumer responded and said, "Well, this is something new, give us more of it."'

The point is this. Even if there are some objective criteria that make one artwork better than another, as long as context plays a role in our aesthetic appreciation of art, it's not possible to create a tangible measure for aesthetic quality that works for all places in all times. Whatever statistical techniques, or artificial intelligence tricks, or machine-learning algorithms you deploy, trying to use numbers to latch on to the essence of artistic excellence is like clutching at smoke with your hands.

But an algorithm needs *something* to go on. So, once you take away popularity and inherent quality, you're left with the only thing that can be quantified: a metric for similarity to whatever has gone before.

There's still a great deal that can be done using measures of similarity. When it comes to building a recommendation engine, like the ones found in Netflix and Spotify, similarity is arguably the ideal measure. Both companies have a way to help users discover new films and songs, and, as subscription services, both have an incentive to accurately predict what users will enjoy. They can't base their algorithms on what's popular, or users would just get bombarded with suggestions for Justin Bieber and *Peppa Pig The Movie*. Nor can they base them on any kind of proxy for quality, such as critical reviews, because if they did the home page would be swamped by arthouse snooze-fests, when all people actually want to do is kick off their shoes after a long day at work and lose themselves in a crappy thriller or stare at Ryan Gosling for two hours.

Similarity, by contrast, allows the algorithm to put the focus squarely on the individual's preferences. What do they listen to, what do they watch, what do they return to time and time again? From there, you can use IMDb or Wikipedia or music blogs or magazine articles to pull out a series of keywords for each song or artist or movie. Do that for the entire catalogue, and then it's a simple step to find and recommend other songs and films with similar tags. Then, in addition, you can find other users who liked similar films and songs, see what other songs and films they enjoyed and recommend those to your user.

At no point is Spotify or Netflix trying to deliver the perfect song or film. They have little interest in perfection. Spotify Discover doesn't promise to hunt out the one band on earth that is destined to align wholly and flawlessly with your taste and mood. The recommendation algorithms merely offer you songs and films that are good enough to insure you against disappointment. They're giving you an inoffensive way of passing the time. Every now and then they will come up with something that you absolutely love, but it's a bit like cold reading in that sense. You only need a strike every now and then to feel the serendipity of discovering new music. The engines don't need to be right all the time.

Similarity works perfectly well for recommendation engines. But when you ask algorithms to create art without a pure measure for quality, that's where things start to get interesting. Can an algorithm be creative if its only sense of art is what happened in the past?

Good artists borrow; great artists steal – Pablo Picasso

In October 1997, an audience arrived at the University of Oregon to be treated to a rather unusual concert. A lone piano sat on the stage at the front. Then the pianist Winifred Kerner took her place at the keys, poised to play three short separate pieces.

One was a lesser-known keyboard composition penned by the master of the baroque, Johann Sebastian Bach. A second was composed in the style of Bach by Steve Larson, a professor of music at the university. And a third was composed by an algorithm, deliberately designed to imitate the style of Bach.

After hearing the three performances, the audience were asked to guess which was which. To Steve Larson's dismay, the majority voted that his was the piece that had been composed by the computer.

And to collective gasps of delighted horror, the audience were told the music they'd voted as genuine Bach was nothing more than the work of a machine.

Larson wasn't happy. In an interview with the *New York Times* soon after the experiment, he said: 'My admiration for [Bach's] music is deep and cosmic. That people could be duped by a computer program was very disconcerting.'

He wasn't alone in his discomfort. David Cope, the man who created the remarkable algorithm behind the computer composition, had seen this reaction before. 'I [first] played what I called the "game" with individuals,' he told me. 'And when they got it wrong they got angry. They were mad enough at me for just bringing up the whole concept. Because creativity is considered a human endeavour.'[19]

This had certainly been the opinion of David Hofstadter, the cognitive scientist and author who had organized the concert in the first place. A few years earlier, in his 1979 Pulitzer Prize winning book *Gödel, Escher, Bach*, Hofstadter had taken a firm stance on the matter:

> Music is a language of emotions, and until programs have emotions as complex as ours, there is no way a program will write anything beautiful ... To think that we might be able to command a pre-programmed 'music box' to bring forth pieces which Bach might have written is a grotesque and shameful mis-estimation of the depth of the human spirit.[20]

But after hearing the output of Cope's algorithm – the so-called 'Experiments in Musical Intelligence' (EMI) – Hofstadter conceded that perhaps things weren't quite so straightforward: 'I find myself baffled and troubled by EMI,' he confessed in the days following the University of Oregon experiment. 'The only comfort I could take at this point comes from realizing that EMI doesn't generate style on its own. It depends on mimicking prior composers. But that is still not all that much comfort. To my absolute devastation [perhaps] music is much less than I ever thought it was.'[21]

So which is it? Is aesthetic excellence the sole preserve of human endeavour? Or can an algorithm create art? And if an audience couldn't distinguish EMI's music from that of a great master, had this machine demonstrated the capacity for true creativity?

Let's try and tackle those questions in turn, starting with the last one. To form an educated opinion, it's worth pausing briefly to understand how the algorithm works.* Something David Cope was generous enough to explain to me.

The first step in building the algorithm was to translate Bach's music into something that can be understood by a machine: 'You have to place into a database five representations of a single note: the on time, the duration, pitch, loudness and instrument.' For each note in Bach's back catalogue, Cope had to painstakingly enter these five numbers into a computer by hand. There were 371 Bach chorales alone, many harmonies, tens of thousands of notes, five numbers per note. It required a monumental effort from Cope: 'For months, all I was doing every day was typing in numbers. But I'm a person who is nothing but obsessive.'

From there, Cope's analysis took each beat in Bach's music and examined what happened next. For every note that is played in a Bach chorale, Cope made a record of the next note. He stored everything together in a kind of dictionary – a bank in which the algorithm could look up a single chord and find an exhaustive list of all the different places Bach's quill had sent the music next.

In that sense, EMI has some similarities to the predictive text algorithms you'll find on your smartphone. Based on the sentences you've written in the past, the phone keeps a dictionary of the words

* I'd also thoroughly recommend looking up some of Cope's music online. I think the orchestra piece in the style of Vivaldi is my favourite: https://www.youtube.com/watch?v=2kuY3BrmTfQ.

you're likely to want to type next and brings them up as suggestions as you're writing.*

The final step was to let the machine loose. Cope would seed the system with an initial chord and instruct the algorithm to look it up in the dictionary to decide what to play next, by selecting the new chord at random from the list. Then the algorithm repeats the process – looking up each subsequent chord in the dictionary to choose the next notes to play. The result was an entirely original composition that sounds just like Bach himself.†

Or maybe it *is* Bach himself. That's Cope's view, anyway. 'Bach created all of the chords. It's like taking Parmesan cheese and putting it through the grater, and then trying to put it back together again. It would still turn out to be Parmesan cheese.'

Regardless of who deserves the ultimate credit, there's one thing that is in no doubt. However beautiful EMI's music may sound, it is based on a pure recombination of existing work. It's mimicking the patterns found in Bach's music, rather than actually composing any music itself.

More recently, other algorithms have been created that make aesthetically pleasing music that is a step on from pure recombination. One particularly successful approach has been genetic algorithms – another type of machine learning, which tries to exploit the way natural selection works. After all, if peacocks are anything to go by, evolution knows a thing or two about creating beauty.

---

* You can also use predictive text to 'compose' some text in your own style. Just open a new note and seed the algorithm with a few words to kick you off, like 'I was born'. Then just repeatedly hit the words that pop up on the screen. Here's mine (genuinely). Starts off fine, gets a bit more stressed towards the end: '(I was born) to be a good person and I would be happy to be with you a lot of people I know that you are not getting my emails and I don't have any time for that.'

† Cope also found it was necessary to keep track of a number of other measures as the algorithm went along. For instance, the length of a phrase and the length of a piece proved to be integral to a Bach-like output.

The idea is simple. Within these algorithms, notes are treated like the DNA of music. It all starts with an initial population of 'songs' – each a random jumble of notes stitched together. Over many generations, the algorithm breeds from the songs, finding and rewarding 'beautiful' features within the music to breed 'better' and better compositions as time goes on. I say 'beautiful' and 'better', but – of course – as we already know, there's no way to decide what either of those words means definitively. The algorithm can create poems and paintings as well as music, but – still – all it has to go on is a measure of similarity to whatever has gone before.

And sometimes that's all you need. If you're looking for a background track for your website or your YouTube video that sounds generically like a folk song, you don't care that it's similar to all the best folk songs of the past. Really, you just want something that avoids copyright infringement without the hassle of having to compose it yourself. And if that's what you're after, there are a number of companies who can help. British startups Jukebox and AI Music are already offering this kind of service, using algorithms that are capable of creating music. Some of that music will be useful. Some of it will be (sort of) original. Some of it will be beautiful, even. The algorithms are undoubtedly great imitators, just not very good innovators.

That's not to do these algorithms a disservice. Most human-made music isn't particularly innovative either. If you ask Armand Leroi, the evolutionary biologist who studied the cultural evolution of pop music, we're a bit too misty-eyed about the inventive capacities of humans. Even the stand-out successes in the charts, he says, could be generated by a machine. Here's his take on Pharrell Williams' 'Happy', for example (something tells me he's not a fan):

'Happy, happy, happy, I'm so happy.' I mean, really! It's got about, like, five words in the lyrics. It's about as robotic a song as you could possibly get, which panders to just the most base uplifting human desire for uplifting summer happy music. The most moronic and reductive song passable. And if that's the level – well it's not too hard.

Leroi doesn't think much of the lyrical prowess of Adele either: 'If you were to analyse any of the songs you would find no sentiment in there that couldn't be created by a sad song generator.'

You may not agree (I'm not sure I do), but there is certainly an argument that much of human creativity – like the products of the 'composing' algorithms – is just a novel combination of pre-existing ideas. As Mark Twain says:

> There is no such thing as a new idea. It is impossible. We simply take a lot of old ideas and put them into a sort of mental kaleidoscope. We give them a turn and they make new and curious combinations. We keep on turning and making new combinations indefinitely; but they are the same old pieces of colored glass that have been in use through all the ages.[22]

Cope, meanwhile, has a very simple definition for creativity, which easily encapsulates what the algorithms can do: 'Creativity is just finding an association between two things which ordinarily would not seem related.'

Perhaps. But I can't help feeling that if EMI and algorithms like it *are* exhibiting creativity, then it's a rather feeble form. Their music might be beautiful, but it is not profound. And try as I might, I can't quite shake the feeling that seeing the output of these machines as art leaves us with a rather culturally impoverished view of the world. It's cultural comfort food, maybe. But not art with a capital A.

In researching this chapter, I've come to realize that the source of my discomfort about algorithms making art lies in a different question. The real issue is not whether machines can be creative. They can. It is about what counts as art in the first place.

I'm a mathematician. I can trade in facts about false positives and absolute truths about accuracy and statistics with complete confidence. But in the artistic sphere I'd prefer to defer to Leo Tolstoy. Like him, I think that true art is about human connection; about communicating emotion. As he put it: 'Art is not a handicraft, it is the transmission of feeling the artist has experienced.'[23] If you agree with Tolstoy's argument then there's a reason why machines can't produce true art. A reason expressed beautifully by Douglas Hofstadter, years before he encountered EMI:

> A 'program' which could produce music ... would have to wander around the world on its own, fighting its way through the maze of life and feeling every moment of it. It would have to understand the joy and loneliness of a chilly night wind, the longing for a cherished hand, the inaccessibility of a distant town, the heartbreak and regeneration after a human death. It would have to have known resignation and world-weariness, grief and despair, determination and victory, piety and awe. It would have had to commingle such opposites as hope and fear, anguish and jubilation, serenity and suspense. Part and parcel of it would have to be a sense of grace, humour, rhythm, a sense of the unexpected – and of course an exquisite awareness of the magic of fresh creation. Therein, and only therein, lie the sources of meaning in music.[24]

I might well be wrong here. Perhaps if algorithmic art takes on the appearance of being a genuine human creation – as EMI did – we'll still value it, and bring our own meaning to it. After all, the long history of manufactured pop music seems to hint that humans can form an emotional reaction to something that has no more than the semblance of an authentic connection. And perhaps once these algorithmic artworks become more commonplace and we become aware that the art didn't come from a human, we won't be bothered by the one-way connection. After all, people form emotional relationships with objects that don't love them back – like treasured childhood teddy bears or pet spiders.

But for me, true art can't be created by accident. There are boundaries to the reach of algorithms. Limits to what can be quantified. Among all of the staggeringly impressive, mind-boggling things that data and statistics can tell me, how it feels to be human isn't one of them.

Conclusion

Rahinah Ibrahim was an architect with four children, a husband who lived overseas, a job volunteering at a local hospital and a PhD at Stanford to complete. As if her life wasn't busy enough, she had also just undergone an emergency hysterectomy and – although she was pretty much back on her feet by now – was still struggling with standing unaided for any length of time without medication. None the less, when the 38th annual International Conference on System Sciences rolled around in January 2005 she booked her flights to Hawaii and organized herself to present her latest paper to her academic peers.[1]

When Ibrahim arrived at San Francisco airport with her daughter, first thing on the morning of 2 January 2005, she approached the counter, handed over her documents and asked the staff if they could help her source some wheelchair assistance. They did not oblige. Her name flashed up on the computer screen as belonging to the federal no-fly list – a database set up after 9/11 to prevent suspected terrorists from travelling.

Ibrahim's teenage daughter, left alone and distraught by the desk, called a family friend saying they'd marched her mother away in handcuffs. Ibrahim, meanwhile, was put into the back of a police car and taken to the station. They searched beneath her hijab, refused her medication and locked her in a cell. Two hours later a Homeland Security agent arrived with release papers and told her she had been taken off the list. Ibrahim made it to her conference in Hawaii and then flew on to her native Malaysia to visit family.

Ibrahim had been put on the no-fly list when an FBI agent ticked the wrong box on a form. It might be that the mistake was down to a mix-up between Jemaah Islamiyah, a terrorist organization notorious for the Bali bombings of 2002, and Jemaah Islam, a professional Malaysian organization for people who study abroad. Ibrahim was a member of the latter, but had never had any connection with the former. It was a simple mistake, but one with dramatic consequences. As soon as the error had made its way into the automated system, it had taken on an aura of authority that made it all but immune to appeal. The encounter at San Francisco wasn't the end of the story.

On the return leg of her journey two months later, while flying home to the United States from Malaysia, Ibrahim was again stopped at the airport. This time, the resolution did not come so quickly. Her visa had been revoked on the grounds of suspected connections to terrorism. Although she was the mother of an American citizen, had her home in San Francisco and held a role at one of the country's most prestigious universities, Ibrahim was not allowed to return to the United States. In the end, it would take almost a decade of fighting to win the case to clear her name. Almost a decade during which she was forbidden to set foot on American soil. And all because of one human error, and a machine with an omnipotent authority.

Human plus machine

There's no doubting the profound positive impact that automation has had on all of our lives. The algorithms we've built to date boast a bewilderingly impressive list of accomplishments. They can help us help diagnose breast cancer, catch serial killers and avoid plane crashes; give each of us free and easy access to the full wealth of human knowledge with our fingertips; and connect people across the globe instantly in a way that our ancestors could only have dreamed of. But in our urge to automate, in our hurry to solve many of the world's issues, we seem to have swapped one problem for another. The algorithms – useful and impressive as they are – have left us with a tangle of complications to unpick.

Everywhere you look – in the judicial system, in healthcare, in policing, even online shopping – there are problems with privacy, bias, error, accountability and transparency that aren't going to go away easily. Just by virtue of some algorithms existing, we face issues of fairness that cut to the core of who we are as humans, what we want our society to look like, and how far we can cope with the impending authority of dispassionate technology.

But maybe that's precisely the point. Perhaps thinking of algorithms as some kind of authority is exactly where we're going wrong.

For one thing, our reluctance to question the power of an algorithm has opened the door to people who wish to exploit us. In researching this book, I have come across all manner of snake-oil salesmen willing to trade on myths and profit from our gullibility. Despite the weight of scientific evidence to the contrary, there are people selling algorithms to police forces and governments that claim to 'predict' whether someone is a terrorist or a paedophile based on the characteristics of their face alone. Others insist their algorithm can suggest changes to a single line in a screenplay that will make a movie more profitable at the box office.* Others boldly

---

* I actually met the CEO of this particular company for an interview. I asked him if he'd ever validated his algorithms to see if they lived up to the claims, and he launched into a long anecdote about how the neural network's analysis had led to a major Hollywood star being dropped from a movie franchise. When I pointed out that was evidence that people bought into his algorithm, not that the algorithm worked, he said: 'Well, we're not running an academic exercise.'

state – without even a hint of sarcasm – that their algorithm is capable of finding your one true love.*

But even the algorithms that live up to their claims often misuse their authority. This book is packed full of stories of the harm that algorithms can do. The 'budget tool' used to arbitrarily cut financial assistance to disabled residents of Idaho. The recidivism algorithms that, thanks to historical data, are more likely to suggest a higher risk score for black defendants. The kidney injury detection system that forces millions of people to give up their most personal and private data without their consent or knowledge. The supermarket algorithm that robs a teenage girl of the chance to tell her father that she's fallen pregnant. The Strategic Subject List that was intended to help victims of gun crime, but was used by police as a hit list. Examples of unfairness are everywhere.

And yet, pointing out the flaws in the algorithms risks implying that there is a perfect alternative we're aiming for. I've thought long and hard and I've struggled to find a single example of a perfectly fair algorithm. Even the ones that look good on the surface – like autopilot in planes or neural networks that diagnose cancer – have problems deep down. As you'll have read in the 'Cars' chapter, autopilot can put those who trained under automation at a serious disadvantage behind the wheel or the joystick. There are even concerns that the apparently miraculous tumour-finding algorithms we looked at in the 'Medicine' chapter don't work as well on all ethnic groups. But examples of perfectly fair, just systems aren't exactly

---

* There's a trick you can use to spot the junk algorithms. I like to call it the Magic Test. Whenever you see a story about an algorithm, see if you can swap out any of the buzzwords, like 'machine learning', 'artificial intelligence' and 'neural network', and swap in the word 'magic'. Does everything still make grammatical sense? Is any of the meaning lost? If not, I'd be worried that something smells quite a lot like bullshit. Because I'm afraid – long into the foreseeable future – we're not going to 'solve world hunger with magic' or 'use magic to write the perfect screenplay' any more than we are with AI.

abundant when algorithms *aren't* involved either. Wherever you look, in whatever sphere you examine, if you delve deep enough into any system at all, you'll find some kind of bias.

So, imagine for a moment: what if we accepted that perfection doesn't exist? Algorithms *will* make mistakes. Algorithms *will* be unfair. That should in no way distract us from the fight to make them more accurate and less biased wherever we can – but perhaps acknowledging that algorithms aren't perfect, any more than humans are, might just have the effect of diminishing any assumption of their authority.

Imagine that, rather than exclusively focusing our attention on designing our algorithms to adhere to some impossible standard of perfect fairness, we instead designed them to facilitate redress when they inevitably erred; that we put as much time and effort into ensuring that automatic systems were as easy to challenge as they are to implement. Perhaps the answer is to build algorithms to be contestable from the ground up. Imagine that we designed them to support humans in their decisions, rather than instruct them. To be transparent about why they came to a particular decision, rather than just inform us of the result.

In my view, the best algorithms are the ones that take the human into account at every stage. The ones that recognize our habit of over-trusting the output of a machine, while embracing their own flaws and wearing their uncertainty proudly front and centre.

This was one of the best features of the IBM Watson *Jeopardy*-winning machine. While the format of the quiz show meant it had to commit to a single answer, the algorithm also presented a series of alternatives it had considered in the process, along with a score indicating how confident it was in each being correct. Perhaps if likelihood of recidivism scores included something similar, judges might find it easier to question the information the algorithm was offering. And perhaps if facial recognition algorithms presented a number of possible matches, rather than just homing in on a single face, misidentification might be less of an issue.

The same feature is what makes the neural networks that screen breast cancer slides so effective. The algorithm doesn't dictate which patients have tumours. It narrows down the vast array of cells to a handful of suspicious areas for the pathologist to check. The algorithm never gets tired and the pathologist rarely misdiagnoses. The algorithm and the human work together in partnership, exploiting each other's strengths and embracing each other's flaws.

There are other examples, too – including in the world of chess, where this book began. Since losing to Deep Blue, Garry Kasparov hasn't turned his back on computers. Quite the opposite. Instead, he has become a great advocate of the idea of 'Centaur Chess', where a human player and an algorithm collaborate with one another to compete with another hybrid team. The algorithm assesses the possible consequences of each move, reducing the chance of a blunder, while the human remains in charge of the game.

Here's how Kasparov puts it: 'When playing with the assistance of computers, we could concentrate on strategic planning instead of spending so much time on calculations. Human creativity was even more paramount under these conditions.'[2] The result is chess played at a higher level than has ever been seen before. Perfect tactical play and beautiful, meaningful strategies. The very best of both worlds.

This is the future I'm hoping for. One where the arrogant, dictatorial algorithms that fill many of these pages are a thing of the past. One where we stop seeing machines as objective masters and start treating them as we would any other source of power. By questioning their decisions; scrutinizing their motives; acknowledging our emotions; demanding to know who stands to benefit; holding them accountable for their mistakes; and refusing to become complacent. I think this is the key to a future where the net over-all effect of algorithms is a positive force for society. And it's only right that it's a job that rests squarely on our shoulders. Because one thing is for sure. In the age of the algorithm, humans have never been more important.

Acknowledgements

THERE ARE SOME PEOPLE, I imagine, who find writing easy. You know the sort – the ones who jump out of bed before sunrise, have a chapter written by lunch and forget to come down to dinner because they're so at one with their creative flow that they just didn't realize the time.

I am definitely not one of those people.

Getting through this process involved a daily battle with the side of my character that just wants to sit on the sofa eating crisps and watching Netflix, and an all-out war with the tsunamis of worry and panic that I thought I'd left behind when I finished my PhD. I didn't really write this book so much as drag it out of myself, kicking and screaming. Sometimes literally.

So I'm all the more grateful to the remarkable group of people who were willing to help me along the way. My fantastic publishing team, who have been so generous with their time and ideas over the past year: Susanna Wadeson, Quynh Do, Claire Conrad, Emma Parry, Gillian Somerscales, Emma Burton, Sophie Christopher,

Hannah Bright, Caroline Saine and all the people at Janklow and Nesbit, Transworld and Norton who have been helping behind the scenes. Likewise, Sue Rider, Kat Bee and Tom Copson. I'd be lost without you.

Enormous thanks, too, to my interviewees, some of whom are quoted in the text, but all of whom helped shape the ideas for the book: Jonathan Rowson, Nigel Harvey, Adam Benforado, Giles Newell, Richard Berk, Sheena Urwin, Steyve Colgan, Mandeep Dhami, Adrian Weller, Toby Davies, Rob Jenkins, Jon Kanevsky, Timandra Harkness, Dan Popple and the team at West Midlands police, Andy Beck, Jack Stilgoe, Caroline Rance, Paul Newman, Phyllis Illarmi, Armand Leoni, David Cope, Ed Finn, Kate Devlin, Shelia Hayman, Tom Chatwin, Carl Gombrich, Johnny Ryan, Jon Crowcroft and Frank Kelly.

There's also Sue Webb and Debbie Enright from Network Typing and Sharon Richardson, Shruthi Rao and Will Storr, whose help in wrestling this book into shape was invaluable. Plus, once I had finally something approaching sentences written down, James Fulker, Elisabeth Adlington, Brendan Maginnis, Ian Hunter, Omar Miranda, Adam Dennett, Michael Veale, Jocelyn Bailey, Cat Black, Tracy Fry, Adam Rutherford and Thomas Oléron Evans, all helped me find the biggest flaws and beat them into submission. And Geoff Dahl, who, as well as offering moral support throughout this entire process, also had the very clever idea for the cover design.

Very many thanks to my peer reviewers: Elizabeth Cleverdon, Bethany Davies, Ben Dickson, Mike Downes, Charlie and Laura Galan, Katie Heath, Mia Kazi-Fornari, Fatah Ioualitene, Siobhan Mathers, Mabel Smaller, Ali Seyhun Saral, Jennifer Shelley, Edward Steele, Daniel Vesma, Jass Ubhi.

I am also unimaginably grateful to my family, for their unwavering support and steadfast loyalty. Phil, Tracy, Natalie, Marge & Parge, Omar, Mike and Tania – you were more patient with me than

Acknowledgements

I often deserved. (Although don't take that too literally, because I probably am going to write another book, and I need you to help me again, OK?)

And last, but by no means least, Edith. Frankly, you were no help whatsoever, but I wouldn't have had it any other way.

Photograph Credits

1. Page 12: 'Car–dog', reproduced by permission of Danilo Vasconcellos Vargas, Kyushu University, Fukuoka, Japan.
2. Page 89: 'Gorilla in chest', reproduced by permission of Trafton Drew, University of Utah, Salt Lake City, USA.
3. Page 160: 'Steve Talley images' © Steve Talley (left) and FBI.
4. Page 162: 'Neil Douglas and doppelgänger', reproduced by permission of Neil Douglas.
5. Page 167: 'Tortoiseshell glasses', reproduced by permission of Mahmood Sharif, Carnegie Mellon University, Pittsburgh, USA; 'Milla Jovovich at the Cannes Film Festival' by Georges Biard.

Notes

A note on the title

6. Brian W. Kernighan and Dennis M. Ritchie. *The C Programming Language* (Upper Saddle River, NJ: Prentice-Hall, 1978).

Introduction

1. Robert A. Caro, *The Power Broker: Robert Moses and the Fall of New York* (London: Bodley Head, 2015), p. 318.
2. There are a couple of fantastic essays on this very idea that are well worth reading. First, Langdon Winner, 'Do artifacts have politics?', *Daedalus*, vol. 109, no. 1, 1980, pp. 121–36, https://www.jstor.org/stable/20024652, which includes the example of Moses' bridges. And a more modern version: Kate Crawford, 'Can an algorithm be agonistic? Ten scenes from life in calculated publics', *Science, Technology and Human Values*, vol. 41, no. 1, 2016, pp. 77–92.
3. *Scunthorpe Evening Telegraph*, 9 April 1996.
4. Chukwuemeka Afigbo (@nke_ise) posted a short video of this effect on Twitter. Worth looking up if you haven't seen it. It's also on YouTube: https://www.youtube.com/watch?v=87QwWpzVy7I.
5. CNN interview, Mark Zuckerberg: 'I'm really sorry that this happened', *YouTube*, 21 March 2018, https://www.youtube.com/watch?v=G6DOhioBfyY.

Power

1. From a private conversation with the chess grandmaster Jonathan Rowson.
2. Feng-Hsiung Hsu, 'IBM's Deep Blue Chess grandmaster chips', *IEEE Micro*, vol. 19, no. 2, 1999, pp. 70–81, http://ieeexplore.ieee.org/document/755469/.
3. Garry Kasparov, *Deep Thinking: Where Machine Intelligence Ends and Human Creativity Begins* (London: Hodder & Stoughton, 2017).
4. TheGoodKnight, 'Deep Blue vs Garry Kasparov Game 2 (1997 Match)', *YouTube*, 18 Oct. 2012, https://www.youtube.com/watch?v=3Bd1Q2rOmok&t=2290s.
5. Ibid.
6. Steven Levy, 'Big Blue's Hand of God', *Newsweek*, 18 May 1997, http://www.newsweek.com/big-blues-hand-god-173076.
7. Kasparov, *Deep Thinking*, p. 187.
8. Ibid., p. 191.
9. According to *Merriam–Webster*. The *Oxford English Dictionary*'s definition makes more of the mathematical nature of algorithms: 'a process or set of rules to be followed in calculations or other problem-solving operations, especially by a computer'.
10. There are lots of different ways you could group algorithms, and I have no doubt that some computer scientists will complain that this list is too simplistic. It's true that a more exhaustive list would have included several other category headers: mapping, reduction, regression and clustering, to name a few. But in the end, I chose this particular set of categories – from Nicholas Diakopoulos, *Algorithmic Accountability Reporting: On the Investigation of Black Boxes* (New York: Tow Center for Digital Journalism, Columbia University, 2014) – as it does a great job at covering the basics and offers a useful way to demystify and distil a vast, complex area of study.
11. Kerbobotat, 'Went to buy a baseball bat on Amazon, they have some interesting suggestions for accessories', *Reddit*, 28 Sept. 2013, https://www.reddit.com/r/funny/comments/1nb16l/went_to_buy_a_baseball_bat_on_amazon_they_have/.
12. Sarah Perez, 'Uber debuts a "smarter" UberPool in Manhattan', *TechCrunch*, 22 May 2017, https://techcrunch.com/2017/05/22/uber-debuts-a-smarter-uberpool-in-manhattan/.
13. I say 'in theory' deliberately. The reality might be a little different. Some algorithms have been built over years by hundreds, even thousands, of developers, each incrementally adding their own steps to the process. As the lines of code grow, so does the complexity of the system, until the logical threads become like a tangled plate of spaghetti. Eventually, the algorithm becomes impossible to follow, and far too complicated for any one human to understand.

    In 2013 Toyota was ordered to pay $3 million in compensation after a fatal crash involving one of its vehicles. The car had accelerated uncontrollably, despite the driver having her foot on the brake rather than the throttle at the time. An expert witness told the jury that an unintended instruction, hidden deep within the vast tangled mess of software, was to blame. See Phil Koopman, *A case study of Toyota unintended acceleration and software safety* (Pittsburgh: Carnegie Mellon University, 18 Sept. 2014), https://users.ece.cmu.edu/~koopman/pubs/koopman14_toyota_ua_slides.pdf.

14. This illusion (the example here is taken from https://commons.wikimedia.org/wiki/File:Vase_of_rubin.png) is known as Rubin's vase, after Edgar Rubin, who developed the idea. It is an example of an *ambiguous* image – right on the border between two shadowed faces looking towards each other, and an image of a white vase. As it's drawn, it's fairly easy to switch between the two in your mind, but all it would take is a couple of lines on the picture to push it in one direction or another. Perhaps a faint outline of the eyes on the faces, or the shadow on the neck of the vase.

The dog/car example of image recognition is a similar story. The team found a picture that was right on the cusp between two different classifications and used the smallest perturbation possible to shift the image from one category to another in the eye of the machine.

15. Jiawei Su, Danilo Vasconcellos Vargas and Kouichi Sakurai, 'One pixel attack for fooling deep neural networks', *arXiv:1719.08864v4* [cs.LG], 22 Feb. 2018, https://arxiv.org/pdf/1710.08864.pdf.

16. Chris Brooke, '"I was only following satnav orders" is no defence: driver who ended up teetering on cliff edge convicted of careless driving', *Daily Mail*, 16 Sept. 2009, http://www.dailymail.co.uk/news/article-1213891/Driver-ended-teetering-cliff-edge-guilty-blindly-following-sat-nav-directions.html#ixzz59vihbQ2n.

17. Ibid.

18. Robert Epstein and Ronald E. Robertson, 'The search engine manipulation effect (SEME) and its possible impact on the outcomes of elections', *Proceedings of the National Academy of Sciences*, vol. 112, no. 33, 2015, pp. E4512–21, http://www.pnas.org/content/112/33/E4512.

19. David Shultz, 'Internet search engines may be influencing elections', *Science*, 7 Aug. 2015, http://www.sciencemag.org/news/2015/08/internet-search-engines-may-be-influencing-elections.

20. Epstein and Robertson, 'The search engine manipulation effect (SEME)'.

21. Linda J. Skitka, Kathleen Mosier and Mark D. Burdick, 'Accountability and automation bias', *International Journal of Human–Computer Studies*, vol. 52, 2000, pp. 701–17, http://lskitka.people.uic.edu/IJHCS2000.pdf.

22. KW *v.* Armstrong, US District Court, D. Idaho, 2 May 2012, https://scholar.google.co.uk/scholar_case?case=17062168494596747089&hl=en&as_sdt=2006.

23. Jay Stanley, *Pitfalls of Artificial Intelligence Decision making Highlighted in Idaho ACLU Case*, American Civil Liberties Union, 2 June 2017, https://www.aclu.org/blog/privacy-technology/pitfalls-artificial-intelligence-decisionmaking-highlighted-idaho-aclu-case.

24. 'K.W. v. Armstrong', *Leagle.com*, 24 March 2014, https://www.leagle.com/decision/infdco20140326c20.

25. Ibid.

26. ACLU Idaho staff, https://www.acluidaho.org/en/about/staff.

27. Stanley, *Pitfalls of Artificial Intelligence Decision-making*.

28. ACLU, *Ruling mandates important protections for due process rights of Idahoans with developmental disabilities*, 30 March 2016, https://www.aclu.org/news/federal-court-rules-against-idaho-department-health-and-welfare-medicaid-class-action.

29. Stanley, *Pitfalls of Artificial Intelligence Decision-making*.

30. Ibid.

31. Ibid.

32. Ibid.

33. Ibid.

34. Kristine Phillips, 'The former Soviet officer who trusted his gut – and averted a global nuclear catastrophe', *Washington Post*, 18 Sept. 2017, https://www.washingtonpost.com/news/retropolis/wp/2017/09/18/the-former-soviet-officer-who-trusted-his-gut-and-averted-a-global-nuclear-catastrophe/?utm_term=.6546e0f06cce.

35. Pavel Aksenov, 'Stanislav Petrov: the man who may have saved the world', BBC News, 26 Sept. 2013, http://www.bbc.co.uk/news/world-europe-24280831.

36. Ibid.

37. Stephen Flanagan, *Re: Accident at Smiler Rollercoaster, Alton Towers, 2 June 2015: Expert's Report*, prepared at the request of the Health and Safety Executive, Oct. 2015, http://www.chiark.greenend.org.uk/~ijackson/2016/Expert%20witness%20report%20from%20Steven%20Flanagan.pdf.

38. Paul E. Meehl, *Clinical versus Statistical Prediction: A Theoretical Analysis and a Review of the Evidence* (Minneapolis: University of Minnesota, 1996; first publ. 1954), http://citeseerx.ist.psu.edu/viewdoc/download?doi=10.1.1.693.6031&rep=rep1&type=pdf.

39. William M. Grove, David H. Zald, Boyd S. Lebow, Beth E. Snitz and Chad Nelson, 'Clinical versus mechanical prediction: a meta-analysis', *Psychological Assessment*, vol. 12, no. 1, 2000, p. 19.

40. Berkeley J. Dietvorst, Joseph P. Simmons and Cade Massey. 'Algorithmic aversion: people erroneously avoid algorithms after seeing them err', *Journal of Experimental Psychology*, Sept. 2014, http://opim.wharton.upenn.edu/risk/library/WPAF201410-AlgorithmAversion-Dietvorst-Simmons-Massey.pdf.

Data

1. Nicholas Carlson, 'Well, these new Zuckerberg IMs won't help Facebook's privacy problems', *Business Insider*, 13 May 2010, http://www.businessinsider.com/well-these-new-zuckerberg-ims-wont-help-facebooks-privacy-problems-2010-5?IR=T.

2. Clive Humby, Terry Hunt and Tim Phillips, *Scoring Points: How Tesco Continues to Win Customer Loyalty* (London: Kogan Page, 2008).

3. Ibid., Kindle edn, 1313–17.

4. See Eric Schmidt, 'The creepy line', *YouTube*, 11 Feb. 2013, https://www.youtube.com/watch?v=o-rvER6YTss.

5. Charles Duhigg, 'How companies learn your secrets', *New York Times*, 16 Feb. 2012, https://www.nytimes.com/2012/02/19/magazine/shopping-habits.html.

6. Ibid.

7. Sarah Buhr, 'Palantir has raised $880 million at a $20 billion valuation', *TechCrunch*, 23 Dec. 2015.

8. Federal Trade Commission, *Data Brokers: A Call for Transparency and Accountability*, (Washington DC, May 2014), https://www.ftc.gov/system/files/documents/reports/data-brokers-call-transparency-accountability-report-federal-trade-commission-may-2014/140527databrokerreport.pdf.

9. Ibid.

10. Wolfie Christl, *Corporate Surveillance in Everyday Life*, Cracked Labs, June 2017, http://crackedlabs.org/en/corporate-surveillance.

11. Heidi Waterhouse, 'The death of data: retention, rot, and risk', The Lead Developer, Austin, Texas, 2 March 2018, https://www.youtube.com/watch?v=mXvPChEo9iU.

12. Amit Datta, Michael Carl Tschantz and Anupam Datta, 'Automated experiments on ad privacy settings', *Proceedings on Privacy Enhancing Technologies*, no. 1, 2015, pp. 92–112.

13. Latanya Sweeney, 'Discrimination in online ad delivery', *Queue*, vol. 11, no. 3, 2013, p. 10, https://dl.acm.org/citation.cfm?id=2460278.

14. Jon Brodkin, 'Senate votes to let ISPs sell your Web browsing history to advertisers', *Ars Technica*, 23 March 2017, https://arstechnica.com/tech-policy/2017/03/senate-votes-to-let-isps-sell-your-web-browsing-history-to-advertisers/.

15. Svea Eckert and Andreas Dewes, 'Dark data', DEFCON Conference 25, 20 Oct. 2017, https://www.youtube.com/watch?v=1nvYGi7-Lxo.

16. The researchers based this part of their work on Arvind Narayanan and Vitaly Shmatikov, 'Robust de-anonymization of large sparse datasets', paper presented to IEEE Symposium on Security and Privacy, 18–22 May 2008.

17. Michal Kosinski, David Stillwell and Thore Graepel. 'Private traits and attributes are predictable from digital records of human behavior', vol. 110, no. 15, 2013, pp. 5802–5.

18. Ibid.

19. Wu Youyou, Michal Kosinski and David Stillwell, 'Computer-based personality judgments are more accurate than those made by humans', *Proceedings of the National Academy of Sciences*, vol. 112, no. 4, 2015, pp. 1036–40.

20. S. C. Matz, M. Kosinski, G. Nave and D. J. Stillwell, 'Psychological targeting as an effective approach to digital mass persuasion', *Proceedings of the National Academy of Sciences*, vol. 114, no. 48, 2017, 201710966.

21. Paul Lewis and Paul Hilder, 'Leaked: Cambridge Analytica's blueprint for Trump victory', *Guardian*, 23 March 2018.

22. 'Cambridge Analytica planted fake news', BBC, 20 March 2018, http://www.bbc.co.uk/news/av/world-43472347/cambridge-analytica-planted-fake-news.

23. Adam D. I. Kramer, Jamie E. Guillory and Jeffrey T. Hancock, 'Experimental evidence of massive-scale emotional contagion through social networks', *Proceedings of the National Academy of Sciences*, vol. 111, no. 24, 2014, pp. 8788–90.

24. Jamie Bartlett, 'Big data is watching you – and it wants your vote', *Spectator*, 24 March 2018.

25. Li Xiaoxiao, 'Ant Financial Subsidiary Starts Offering Individual Credit Scores', *Caixin*, 2 March 2015, https://www.caixinglobal.com/2015-03-02/101012655.html.

26. Rick Falkvinge, 'In China, your credit score is now affected by your political opinions – and your friends' political opinions', *Privacy News Online*, 3 Oct. 2015, https://www.privateinternetaccess.com/blog/2015/10/in-china-your-credit-score-is-now-affected-by-your-political-opinions-and-your-friends-political-opinions/.

27. *State Council Guiding Opinions Concerning Establishing and Perfecting Incentives for Promise-keeping and Joint Punishment Systems for Trust-breaking, and Accelerating the Construction of Social Sincerity*, China Copyright and Media, 30 May 2016, updated 18 Oct. 2016, https://chinacopyrightandmedia.wordpress.com/2016/05/30/state-council-guiding-opinions-concerning-establishing-and-perfecting-incentives-for-promise-keeping-and-joint-punishment-systems-for-trust-breaking-and-accelerating-the-construction-of-social-sincer/.

28. Rachel Botsman, *Who Can You Trust? How Technology Brought Us Together – and Why It Could Drive Us Apart* (London: Penguin, 2017), Kindle edn, p. 151.

Justice

1. John-Paul Ford Rojas, 'London riots: Lidl water thief jailed for six months', *Telegraph*, 7 Jan. 2018, http://www.telegraph.co.uk/news/uknews/crime/8695988/London-riots-Lidl-water-thief-jailed-for-six-months.html.

2. Matthew Taylor, 'London riots: how a peaceful festival in Brixton turned into a looting free-for-all', *Guardian*, 8 Aug. 2011, https://www.theguardian.com/uk/2011/aug/08/london-riots-festival-brixton-looting.

3. Rojas, 'London riots'.

4. Josh Halliday, 'London riots: how BlackBerry Messenger played a key role', *Guardian*, 8 Aug. 2011, https://www.theguardian.com/media/2011/aug/08/london-riots-facebook-twitter-blackberry.

5. David Mills, 'Paul and Richard Johnson avoid prison over riots', *News Shopper*, 13 Jan. 2012, http://www.newsshopper.co.uk/londonriots/9471288.Father_and_son_avoid_prison_over_riots/.

6. Ibid.

7. Rojas, 'London riots'. 'Normally, the police wouldn't arrest you for such an offence. They wouldn't hold you in custody. They wouldn't take you to court', Hannah Quirk, a senior lecturer in criminal law and justice at Manchester University wrote about Nicholas's case in 2015: Carly Lightowlers and Hannah Quirk, 'The 2011 English "riots": prosecutorial zeal and judicial abandon', *British Journal of Criminology*, vol. 55, no. 1, 2015, pp. 65–85.

8. Mills, 'Paul and Richard Johnson avoid prison over riots'.

9. William Austin and Thomas A. Williams III, 'A survey of judges' responses to simulated legal cases: research note on sentencing disparity', *Journal of Criminal Law and Criminology*, vol. 68, no. 2, 1977, pp. 306–310.

10. Mandeep K. Dhami and Peter Ayton, 'Bailing and jailing the fast and frugal way', *Journal of Behavioral Decision-making*, vol. 14, no. 2, 2001, pp. 141–68, http://onlinelibrary.wiley.com/doi/10.1002/bdm.371/abstract.

11. Up to half of the judges differed in their opinion of the best course of action on any one case.
12. Statisticians have a way to measure this kind of consistency in judgments, called Cohen's Kappa. It takes into account the fact that – even if you were just wildly guessing – you could end up being consistent by chance. A score of one means perfect consistency. A score of zero means you're doing no better than random. The judges' scores ranged from zero to one and averaged 0.69.
13. Diane Machin, 'Sentencing guidelines around the world', paper prepared for Scottish Sentencing Council, May 2005, https://www.scottishsentencingcouncil.org.uk/media/1109/paper-31a-sentencing-guidelines-around-the-world.pdf.
14. Ibid.
15. Ibid.
16. Ernest W. Burgess, 'Factors determining success or failure on parole', in *The Workings of the Intermediate-sentence Law and Parole System in Illinois* (Springfield, IL: State Board of Parole, 1928). It's a tricky paper to track down, so here is an alternative read by Burgess's colleague Tibbitts, on the follow-up study to the original: Clark Tibbitts, 'Success or failure on parole can be predicted: a study of the records of 3,000 youths paroled from the Illinois State Reformatory', *Journal of Criminal Law and Criminology*, vol. 22, no. 1, Spring 1931, https://scholarlycommons.law.northwestern.edu/cgi/viewcontent.cgi?article=2211&context=jclc. The other categories used by Burgess were 'black sheep', 'criminal by accident', 'dope' and 'gangster'. 'Farm boys' were the category he found least likely to re-offend.
17. Karl F. Schuessler, 'Parole prediction: its history and status', *Journal of Criminal Law and Criminology*, vol. 45, no. 4, 1955, pp. 425–31, https://pdfs.semanticscholar.org/4cd2/31dd25321a0c14a9358a93ebccb6f15d3169.pdf.
18. Ibid.
19. Bernard E. Harcourt, *Against Prediction: Profiling, Policing, and Punishing in an Actuarial Age* (Chicago and London: University of Chicago Press, 2007), p. 1.
20. Philip Howard, Brian Francis, Keith Soothill and Les Humphreys, *OGRS 3: The Revised Offender Group Reconviction Scale*, Research Summary 7/09 (London: Ministry of Justice, 2009), https://core.ac.uk/download/pdf/1556521.pdf.
21. A slight caveat here: there probably is some selection bias in this statistic. 'Ask the audience' was typically used in the early rounds of the game, when the questions were a lot easier. None the less, the idea of the collective opinions of a group being more accurate than those of any individual is a well-documented phenomenon. For more on this, see James Surowiecki, *The Wisdom of Crowds: Why the Many Are Smarter than the Few* (New York: Doubleday, 2004), p. 4.
22. Netflix Technology Blog, https://medium.com/netflix-techblog/netflix-recommendations-beyond-the-5-stars-part-2-d9b96aa399f5.
23. Shih-ho Cheng, 'Unboxing the random forest classifier: the threshold distributions', Airbnb Engineering and Data Science, https://medium.com/airbnb-engineering/unboxing-the-random-forest-classifier-the-threshold-distributions-22ea2bb58ea6.
24. Jon Kleinberg, Himabindu Lakkaraju, Jure Leskovec, Jens Ludwig and Sendhil Mullainathan, *Human Decisions and Machine Predictions*, NBER Working Paper no. 23180 (Cambridge, MA: National Bureau of Economic Research, Feb. 2017), http://

www.nber.org/papers/w23180. This study actually used 'gradient boosted decision trees', an algorithm similar to random forests. Both combine the predictions of lots of decision trees to arrive at a decision, but the trees in the gradient-boosted method are grown sequentially, while in random forests they are grown in parallel. To set up this study, the dataset was first chopped in half. One half was used to train the algorithm, the other half was kept to one side. Once the algorithm was ready, it took cases from the half that it had never seen before to try to predict what would happen. (Without splitting the data first, your algorithm would just be a fancy look-up table).

25. Academics have spent time developing statistical techniques to deal with precisely this issue, so that you can still make a meaningful comparison between the respective predictions made by judges and algorithms. For more details on this, see Kleinberg et al., *Human Decisions and Machine Predictions*.

26. 'Costs per place and costs per prisoner by individual prison', *National Offender Management Service Annual Report and Accounts 2015–16*, Management Information Addendum, Ministry of Justice information release, 27 Oct. 2016, https://www.gov.uk/government/uploads/system/uploads/attachment_data/file/563326/costs-per-place-cost-per-prisoner-2015-16.pdf.

27. Marc Santora, 'City's annual cost per inmate is $168,000, study finds', *New York Times*, 23 Aug. 2013, http://www.nytimes.com/2013/08/24/nyregion/citys-annual-cost-per-inmate-is-nearly-168000-study-says.html; Harvard University, 'Harvard at a glance', https://www.harvard.edu/about-harvard/harvard-glance.

28. Luke Dormehl, *The Formula: How Algorithms Solve All Our Problems . . . and Create More* (London: W. H. Allen, 2014), p. 123.

29. Julia Angwin, Jeff Larson, Surya Mattu and Lauren Kirchner, 'Machine bias', ProPublica, 23 May 2016, https://www.propublica.org/article/machine-bias-risk-assessments-in-criminal-sentencing.

30. 'Risk assessment' questionnaire, https://www.documentcloud.org/documents/2702103-Sample-Risk-Assessment-COMPAS-CORE.html.

31. Tim Brennan, William Dieterich and Beate Ehret (Northpointe Institute), 'Evaluating the predictive validity of the COMPAS risk and needs assessment system', *Criminal Justice and Behavior*, vol. 36, no. 1, 2009, pp. 21–40, http://www.northpointeinc.com/files/publications/Criminal-Justice-Behavior-COMPAS.pdf. According to a 2018 study, the COMPAS algorithm has a similar accuracy to an 'ensemble' of humans. The researchers demonstrated that asking a group of 20 inexperienced individuals to predict recidivism achieved an equivalent score to the COMPAS system. It's an interesting comparison, but it's worth remembering that in real courts, judges don't have a team of strangers making votes in the back room. They're on their own. And that's the only comparison that really counts. See Julia Dressel and Hany Farid, 'The accuracy, fairness, and limits of predicting recidivism', *Science Advances*, vol. 4, no. 1, 2018.

32. *Christopher Drew Brooks* v. *Commonwealth*, Court of Appeals of Virginia, Memorandum Opinion by Judge Rudolph Bumgardner III, 28 Jan. 2004, https://law.justia.com/cases/virginia/court-of-appeals-unpublished/2004/2540023.html.

33. 'ACLU brief challenges constitutionality of Virginia's sex offender risk assessment guidelines', American Civil Liberties Union Virginia, 28 Oct. 2003, https://acluva.org/en/press-releases/aclu-brief-challenges-constitutionality-virginias-sex-offender-risk-assessment.

34. *State* v. *Loomis*, Supreme Court of Wisconsin,13 July 2016, http://caselaw.findlaw.com/wi-supreme-court/1742124.html.

35. Quotations from Richard Berk are from personal communication.

36. Angwin et al., 'Machine bias'.

37. *Global Study on Homicide 2013* (Vienna: United Nations Office on Drugs and Crime, 2014), http://www.unodc.org/documents/gsh/pdfs/2014_GLOBAL_HOMICIDE_BOOK_web.pdf.

38. ACLU, 'The war on marijuana in black and white', June 2013, www.aclu.org/files/assets/aclu-the waronmarijuana-ve12.pdf

39. Surprisingly, perhaps, Equivant's stance on this is backed up by the Supreme Court of Wisconsin. After Eric Loomis was sentenced to prison for six years by a judge using the COMPAS risk-assessment tool, he appealed the ruling. The case, *Loomis* v. *Wisconsin*, claimed that the use of proprietary, closed-source risk-assessment software to determine his sentence violated his right to due process, because the defence can't challenge the scientific validity of the score. But the Wisconsin Supreme Court ruled that a trial court's use of an algorithmic risk assessment in sentencing did not violate the defendant's due process rights: Supreme Court of Wisconsin, case no. 2015AP157-CR, opinion filed 13 July 2016, https://www.wicourts.gov/sc/opinion/DisplayDocument.pdf?content=pdf&seqNo=171690.

40. Lucy Ward, 'Why are there so few female maths professors in universities?', *Guardian*, 11 March 2013, https://www.theguardian.com/lifeandstyle/the-womens-blog-with-jane-martinson/2013/mar/11/women-maths-professors-uk-universities.

41. Sonja B. Starr and M. Marit Rehavi, *Racial Disparity in Federal Criminal Charging and Its Sentencing Consequences*, Program in Law and Economics Working Paper no. 12-002 (Ann Arbor: University of Michigan Law School, 7 May 2012), http://economics.ubc.ca/files/2013/05/pdf_paper_marit-rehavi-racial-disparity-federal-criminal.pdf.

42. David Arnold, Will Dobbie and Crystal S. Yang, *Racial Bias in Bail Decisions*, NBER Working Paper no. 23421 (Cambridge, MA: National Bureau of Economic Research, 2017), https://www.princeton.edu/~wdobbie/files/racialbias.pdf.

43. John J. Donohue III, *Capital Punishment in Connecticut, 1973–2007: A Comprehensive Evaluation from 4686 Murders to One Execution* (Stanford, CA, and Cambridge, MA: Stanford Law School and National Bureau of Economic Research, Oct. 2011), https://law.stanford.edu/wp-content/uploads/sites/default/files/publication/259986/doc/slspublic/fulltext.pdf.

44. Adam Benforado, *Unfair: The New Science of Criminal Injustice* (New York: Crown, 2015), p. 197.

45. Sonja B. Starr, *Estimating Gender Disparities in Federal Criminal Cases*, University of Michigan Law and Economics Research Paper no. 12-018 (Ann Arbor: University of Michigan Law School, 29 Aug. 2012), https://ssrn.com/abstract=2144002 or http://dx.doi.org/10.2139/ssrn.2144002.

46. David B. Mustard, 'Racial, ethnic, and gender disparities in sentencing: evidence from the US federal courts', *Journal of Law and Economics*, vol. 44, no. 2, April 2001, pp. 285–314, http://people.terry.uga.edu/mustard/sentencing.pdf.

47. Daniel Kahneman, *Thinking, Fast and Slow* (New York: Farrar, Straus & Giroux, 2011), p. 44.

48. Chris Guthrie, Jeffrey J. Rachlinski and Andrew J. Wistrich, *Blinking on the Bench: How Judges Decide Cases*, paper no. 917 (New York: Cornell University Law Faculty, 2007), http://scholarship.law.cornell.edu/facpub/917.

49. Kahneman, *Thinking, Fast and Slow*, p. 13.
50. Ibid., p. 415,
51. Dhami and Ayton, 'Bailing and jailing the fast and frugal way'.
52. Brian Wansink, Robert J. Kent and Stephen J. Hoch, 'An anchoring and adjustment model of purchase quantity decisions', *Journal of Marketing Research*, vol. 35, 1998, pp. 71–81, http://foodpsychology.cornell.edu/sites/default/files/unmanaged_files/Anchoring-JMR-1998.pdf.
53. Mollie Marti and Roselle Wissler, 'Be careful what you ask for: the effect of anchors on personal injury damages awards', *Journal of Experimental Psychology: Applied*, vol. 6, no. 2, 2000, pp. 91–103.
54. Birte Englich and Thomas Mussweiler, 'Sentencing under uncertainty: anchoring effects in the courtroom', *Journal of Applied Social Psychology*, vol. 31, no. 7, 2001, pp. 1535–51, http://onlinelibrary.wiley.com/doi/10.1111/j.1559-1816.2001.tb02687.x.
55. Birte Englich, Thomas Mussweiler and Fritz Strack, 'Playing dice with criminal sentences: the influence of irrelevant anchors on experts' judicial decision making', *Personality and Social Psychology Bulletin*, vol. 32, 2006, pp. 188–200, https://www.researchgate.net/publication/7389517_Playing_Dice_With_Criminal_Sentences_The_Influence_of_Irrelevant_Anchors_on_Experts%27_Judicial_Decision_Making?enrichId=rgreq-f2fedfeb71aa83f8fad80cc24df3254d-XXX&enrichSource=Y292ZXJQYWdlOzczODk1MTc7QVM6MTAzODIzNjIwMTgyMDIyQDE0MDE3NjQ4ODgzMTA%3D&el=1_x_3&_esc=publicationCoverPdf.
56. Ibid.
57. Ibid.
58. Mandeep K. Dhami, Ian K. Belton, Elizabeth Merrall, Andrew McGrath and Sheila Bird, 'Sentencing in doses: is individualized justice a myth?', under review. Kindly shared through personal communication with Mandeep Dhami.
59. Ibid.
60. Adam N. Glynn and Maya Sen, 'Identifying judicial empathy: does having daughters cause judges to rule for women's issues?', *American Journal of Political Science*, vol. 59, no. 1, 2015, pp. 37–54, https://scholar.harvard.edu/files/msen/files/daughters.pdf.
61. Shai Danziger, Jonathan Levav and Liora Avnaim-Pesso, 'Extraneous factors in judicial decisions', *Proceedings of the National Academy of Sciences of the United States of America*, vol. 108, no. 17, 2011, pp. 6889–92, http://www.pnas.org/content/108/17/6889.
62. Keren Weinshall-Margel and John Shapard, 'Overlooked factors in the analysis of parole decisions', *Proceedings of the National Academy of Sciences of the United States of America*, vol. 108, no. 42, 2011, E833, http://www.pnas.org/content/108/42/E833.long.
63. Uri Simonsohn and Francesca Gino, 'Daily horizons: evidence of narrow bracketing in judgment from 9,000 MBA-admission interviews', *Psychological Science*, vol. 24, no. 2, 2013, pp. 219–24, https://ssrn.com/abstract=2070623.
64. Lawrence E. Williams and John A. Bargh, 'Experiencing physical warmth promotes interpersonal warmth', *Science*, vol. 322, no. 5901, pp. 606–607, https://www.ncbi.nlm.nih.gov/pmc/articles/PMC2737341/.

Medicine

1. Richard M. Levenson, Elizabeth A. Krupinski, Victor M. Navarro and Edward A. Wasserman. 'Pigeons (*Columba livia*) as trainable observers of pathology and radiology breast cancer images', *PLOSOne*, 18 Nov. 2015, http://journals.plos.org/plosone/article?id=10.1371/journal.pone.0141357.
2. 'Hippocrates' daughter as a dragon kills a knight, in "The Travels of Sir John Mandeville"', *British Library Online Gallery*, 26 March 2009, http://www.bl.uk/onlinegallery/onlineex/illmanus/harlmanucoll/h/011hrl000003954u00008v00.html.
3. Eleni Tsiompanou, 'Hippocrates: timeless still', JLL Bulletin: Commentaries on the History of Treatment Evaluation (Oxford and Edinburgh: James Lind Library, 2012), http://www.jameslindlibrary.org/articles/hippocrates-timeless-still/.
4. David K. Osborne, 'Hippocrates: father of medicine', *Greek Medicine.net*, 2015, http://www.greekmedicine.net/whos_who/Hippocrates.html.
5. Richard Colgan, 'Is there room for art in evidence-based medicine?', *AMA Journal of Ethics*, Virtual Mentor 13: 1, Jan. 2011, pp. 52–4, http://journalofethics.ama-assn.org/2011/01/msoc1-1101.html.
6. Joseph Needham, *Science and Civilization in China*, vol. 6, *Biology and Biological Technology*, part VI, *Medicine*, ed. Nathan Sivin (Cambridge: Cambridge University Press, 2004), p. 143, https://monoskop.org/images/1/16/Needham_Joseph_Science_and_Civilisation_in_China_Vol_6-6_Biology_and_Biological_Technology_Medicine.pdf.
7. 'Ignaz Semmelweis', *Brought to Life: Exploring the History of Medicine* (London: Science Museum n.d.), http://broughttolife.sciencemuseum.org.uk/broughttolife/people/ignazsemmelweis.
8. Quotations from Andy Beck are from personal communication.
9. Joann G. Elmore, Gary M. Longton, Patricia A. Carney, Berta M. Geller, Tracy Onega, Anna N. A. Tosteson, Heidi D. Nelson, Margaret S. Pepe, Kimberly H. Allison, Stuart J. Schnitt, Frances P. O'Malley and Donald L. Weaver, 'Diagnostic concordance among pathologists interpreting breast biopsy specimens', *Journal of the American Medical Association*, vol. 313, no. 11, 17 March 2015, 1122–32, https://jamanetwork.com/journals/jama/fullarticle/2203798.
10. Ibid.
11. The name 'neural network' came about as an analogy with what happens in the brain. There, billions of neurons are connected to one another in a gigantic network. Each neuron listens to its connections and sends out a signal whenever it picks up on another neuron being excited. The signal then excites some other neurons that are listening to it.

A neural network is a much simpler and more orderly version of the brain. Its (artificial) neurons are structured in layers, and all the neurons in each layer listen to all the neurons in the previous layer. In our dog example the very first layer is the individual pixels in the image. Then there are several layers with thousands of neurons in them, and a final layer with only a single neuron in it that outputs the probability that the image fed in is a dog.

The procedure for updating the neurons is known as the 'backpropagation algorithm'. We start with the final neuron that outputs the probability that the image is a dog. Let's say we fed in an image of a dog and it predicted that the image had a 70 per cent chance of being a dog. It looks at the signals it received from the previous layer and says, 'The next time I receive information like that I'll increase my probability that the image is a dog'. It then says to each of the neurons in the previous layer, 'Hey, if you'd given me this signal instead I would have made a better prediction'. Each of those neurons looks at its input signals and changes what it would output the next time. And then it tells the previous layer what signals it should have sent, and so on through all the layers back to the beginning. It is this process of propagating the errors back through the neural network that leads to the name 'the backpropagation algorithm'.

For a more detailed overview of neural networks, how they are built and trained, see Pedro Domingos, *The Master Algorithm: How the Quest for the Ultimate Learning Machine Will Remake Our World* (New York: Basic Books, 2015).

12. Alex Krizhevsky, Ilya Sutskever and Geoffrey E. Hinton, 'ImageNet classification with deep convolutional neural networks', in F. Pereira, C. J. C. Burges, L. Bottou and K. Q. Weinberger, eds, *Advances in Neural Information Processing Systems 25* (La Jolla, CA, Neural Information Processing Systems Foundation, 2012), pp. 1097–1105, http://papers.nips.cc/paper/4824-imagenet-classification-with-deep-convolutional-neural-networks.pdf. This particular algorithm is known as a convolutional neural network. Rather than the entire image being fed in, the algorithm first applies a host of different filters and looks for local patterns in the way the picture is distorted.

13. Marco Tulio Ribeiro, Sameer Singh and Carlos Guestrin, '"Why should I trust you?" Explaining the predictions of any classifier', *Proceedings of the 22nd ACM SIGKDD International Conference on Knowledge Discovery and Data Mining*, San Francisco, 2016, pp. 1135–44, http://www.kdd.org/kdd2016/papers/files/rfp0573-ribeiroA.pdf.

14. This was compared to the assessment of a panel of experts, whose collective analysis was considered to be the 'ground truth' for what was contained in the slides.

15. Trafton Drew, Melissa L. H. Vo and Jeremy M. Wolfe, 'The invisible gorilla strikes again: sustained inattentional blindness in expert observers', *Psychological Science*, vol. 24, no. 9, Sept. 2013, pp. 1848–53, https://www.ncbi.nlm.nih.gov/pmc/articles/PMC3964612/.

16. The gorilla is located in the top-right-hand side of the image.

17. Yun Liu, Krishna Gadepalli, Mohammad Norouzi, George E. Dahl, Timo Kohlberger, Aleksey Boyko, Subhashini Venugopalan, Aleksei Timofeev, Philip Q. Nelson, Greg S. Corrado, Jason D. Hipp, Lily Peng and Martin C. Stumpe, 'Detecting cancer metastases on gigapixel pathology images', Cornell University Library, 8 March 2017, https://arxiv.org/abs/1703.02442.

18. Dayong Wang, Aditya Khosla, Rishab Gargeya, Humayun Irshad and Andrew H. Beck, 'Deep learning for identifying metastatic breast cancer', Cornell University Library, 18 June 2016, https://arxiv.org/abs/1606.05718.

19. David A. Snowdon, 'The Nun Study', *Boletin de LAZOS de la Asociación Alzheimer de Monterrey*, vol. 4, no. 22, 2000; D. A. Snowdon, 'Healthy aging and dementia: findings from the Nun Study', *Annals of Internal Medicine*, vol. 139, no. 5, Sept. 2003, pp. 450–54.

20. The idea density – a proxy for linguistic complexity – was calculated by counting up the number of unique ideas each nun used per string of ten words. There's a nice overview here: Associated Press, 'Study of nuns links early verbal skills to Alzheimer's,

Los Angeles Times, 21 Feb. 1996, http://articles.latimes.com/1996-02-21/news/mn-38356_1_alzheimer-nuns-studied.

21. Maja Nielsen, Jørn Jensen and Johan Andersen, 'Pre-cancerous and cancerous breast lesions during lifetime and at autopsy: a study of 83 women', *Cancer*, vol. 54, no. 4, 1984, pp. 612–15, http://onlinelibrary.wiley.com/wol1/doi/10.1002/1097-0142(1984)54:4%3C612::AID-CNCR2820540403%3E3.0.CO;2-B/abstract.

22. H. Gilbert Welch and William C. Black, 'Using autopsy series to estimate the disease "reservoir" for ductal carcinoma in situ of the breast: how much more breast cancer can we find?', *Annals of Internal Medicine*, vol. 127, no. 11, Dec. 1997, pp. 1023–8, www.vaoutcomes.org/papers/Autopsy_Series.pdf.

23. Getting an exact statistic is tricky because it depends on the country and demographic (and how aggressively your country screens for breast cancer). For a good summary, see: http://www.cancerresearchuk.org/health-professional/cancer-statistics/statistics-by-cancer-type/breast-cancer.

24. Quotations from Jonathan Kanevsky are from personal communication.

25. 'Breakthrough method predicts risk of DCIS becoming invasive breast cancer', *Artemis*, May 2010, http://www.hopkinsbreastcenter.org/artemis/201005/3.html.

26. H. Gilbert Welch, Philip C. Prorok, A. James O'Malley and Barnett S. Kramer, 'Breast-cancer tumor size, overdiagnosis, and mammography screening effectiveness', *New England Journal of Medicine*, vol. 375, 2016, pp. 1438–47, http://www.nejm.org/doi/full/10.1056/NEJMoa1600249.

27. Independent UK Panel on Breast Cancer Screening, 'The benefits and harms of breast cancer screening: an independent review', *Lancet*, vol. 380, no. 9855, 30 Oct. 2012, pp. 1778–86, http://www.thelancet.com/journals/lancet/article/PIIS0140-6736(12)61611-0/abstract.

28. Personal communication.

29. Andrew H. Beck, Ankur R. Sangoi, Samuel Leung, Robert J. Marinelli, Torsten O. Nielsen, Marc J. van de Vijver, Robert B. West, Matt van de Rijn and Daphne Koller, 'Systematic analysis of breast cancer morphology uncovers stromal features associated with survival', *Science Transitional Medicine*, 19 Dec. 2014, https://becklab.hms.harvard.edu/files/becklab/files/sci_transl_med-2011-beck-108ra113.pdf.

30. Phi Vu Tran, 'A fully convolutional neural network for cardiac segmentation in short-axis MRI', 27 April 2017, https://arxiv.org/pdf/1604.00494.pdf.

31. 'Emphysema', *Imaging Analytics*, Zebra Medical, https://www.zebra-med.com/algorithms/lungs/.

32. Eun-Jae Lee, Yong-Hwan Kim, Dong-Wha Kang et al., 'Deep into the brain: artificial intelligence in stroke imaging', *Journal of Stroke*, vol. 19, no. 3, 2017, pp. 277–85, https://www.ncbi.nlm.nih.gov/pmc/articles/PMC5647643/.

33. Taylor Kubota, 'Deep learning algorithm does as well as dermatologists in identifying skin cancer', *Stanford News*, 25 Jan. 2017, https://news.stanford.edu/2017/01/25/artificial-intelligence-used-identify-skin-cancer/.

34. Jo Best, 'IBM Watson: the inside story of how the *Jeopardy*-winning supercomputer was born, and what it wants to do next', *Tech Republic*, n.d., https://www.techrepublic.com/article/ibm-watson-the-inside-story-of-how-the-jeopardy-winning-supercomputer-was-born-and-what-it-wants-to-do-next/.

35. Jennings Brown, 'Why everyone is hating on IBM Watson, including the people who helped make it', *Gizmodo*, 14 Aug. 2017, https://www.gizmodo.com.au/2017/08/why-everyone-is-hating-on-watsonincluding-the-people-who-helped-make-it/.

36. https://www.theregister.co.uk/2017/02/20/watson_cancerbusting_trial_on_hold_after_damning_audit_report/

37. Casey Ross and Ike Swetlitz, 'IBM pitched its Watson supercomputer as a revolution in cancer care. It's nowhere close', *STAT*, 5 Sept. 2017, https://www.statnews.com/2017/09/05/watson-ibm-cancer/.

38. Tomoko Otake, 'Big data used for rapid diagnosis of rare leukemia case in Japan', *Japan Times*, 11 Aug. 2016, https://www.japantimes.co.jp/news/2016/08/11/national/science-health/ibm-big-data-used-for-rapid-diagnosis-of-rare-leukemia-case-in-japan/#.Wf8S_hO0MQ8.

39. 'Researchers validate five new genes responsible for ALS', *Science Daily*, 1 Dec. 2017, https://www.sciencedaily.com/releases/2017/12/171201104101.htm.

40. John Freedman, 'A reality check for IBM's AI ambitions', *MIT Technology Review*, 27 June 2017.

41. *Asthma facts and statistics*, Asthma UK, 2016, https://www.asthma.org.uk/about/media/facts-and-statistics/; *Asthma in the US*, Centers for Disease Control and Prevention, May 2011, https://www.cdc.gov/vitalsigns/asthma/index.html.

42. 'Schoolgirl, 13, who died of asthma attack was making regular trips to A&E and running out of medication – but was NEVER referred to a specialist even when her lips turned blue, mother tells inquest', *Daily Mail*, 13 Oct. 2015, http://www.dailymail.co.uk/news/article-3270728/Schoolgirl-13-died-asthma-attack-not-referred-specialist-lips-turned-blue.html.

43. *My Data, My Care: How Better Use of Data Improves Health and Wellbeing* (London: Richmond Group of Charities, Jan. 2017), https://richmondgroupofcharities.org.uk/publications.

44. Terence Carney, 'Regulation 28: report to prevent future deaths', coroner's report on the case of Tamara Mills, 29 Oct. 2015, https://www.judiciary.gov.uk/publications/tamara-mills/.

45. Jamie Grierson and Alex Hern, 'Doctors using Snapchat to send patient scans to each other, panel finds', *Guardian*, 5 July 2017, https://www.theguardian.com/technology/2017/jul/05/doctors-using-snapchat-to-send-patient-scans-to-each-other-panel-finds.

46. Even if you get around all of those issues, sometimes the data itself just doesn't exist. There are thousands of rare diseases with an underlying genetic cause that are effectively unique. Doctors have enormous difficulty spotting one of these conditions because in many cases they will have never seen it before. All the algorithms in the world won't solve issues with tiny sample sizes.

47. Hal Hodson, 'Revealed: Google AI has access to huge haul of NHS patient data', *New Scientist*, 29 April 2016, https://www.newscientist.com/article/2086454-revealed-google-ai-has-access-to-huge-haul-of-nhs-patient-data/.

48. Actually, much of the blame for this so-called 'legally inappropriate' deal has been laid at the door of the Royal Free Trust, which was probably a bit too eager

to partner up with the most famous artificial intelligence company in the world. See the letter from Dame Fiona Caldicott, the national data guardian, that was leaked to Sky News: Alex Martin, 'Google received 1.6 million NHS patients' data on an "inappropriate legal basis"', Sky News, 15 May 2017, https://photos. google.com/share/AF1QipMdd5VTK0RNQ1AC3Dda1526CMG0vPD4P3x4x6_ qmj0Zf101rbKyxfkfyputSPvqdA/photo/AF1QipP1_rnJMXkRyy3IuFHasilQHYEknK gnHFOFEy4T?key= U2pZUDM4bmo5RHhKYVptaDlkbEhfVFh4Rm1iVUVR.

49. Denis Campbell, 'Surgeons attack plans to delay treatment to obese patients and smokers', Guardian, 29 Nov. 2016, https://www.theguardian.com/society/2016/ nov/29/surgeons-nhs-delay-treatment-obese-patients-smokers-york.
50. Nir Eyal, 'Denial of treatment to obese patients: the wrong policy on personal responsibility for health', International Journal of Health Policy and Management, vol. 1, no. 2, Aug. 2013, pp. 107–10, https://www.ncbi.nlm.nih.gov/pmc/articles/ PMC3937915/.
51. For a description of the procedures, see http://galton.org/essays/1880-1889/galton-1884-jaigi-anthro-lab.pdf.
52. Francis Galton, 'On the Anthropometric Laboratory at the late international health exhibition', Journal of the Anthropological Institute of Great Britain and Ireland, vol. 14, 1885, pp. 205–21.
53. 'Taste', https://permalinks.23andme.com/pdf/samplereport_traits.pdf.
54. 'Sneezing on summer solstice?', 23andMeBlog, 20 June 2012, https://blog.23andme. com/health-traits/sneezing-on-summer-solstice/.
55. 'Find out what your DNA says about your health, traits and ancestry', 23andMe, https://www.23andme.com/en-gb/dna-health-ancestry/.
56. Kristen v. Brown, '23andMe is selling your data but not how you think', Gizmodo, 14 April 2017, https://gizmodo.com/23andme-is-selling-your-data-but-not-how-you-think-1794340474.
57. Michael Grothaus, 'How23andMe is monetizing your DNA', Fast Company, 15 Jan. 2015, https://www.fastcompany.com/3040356/what-23andme-is-doing-with-all-that-dna.
58. Rob Stein, 'Found on the Web, with DNA: a boy's father', Washington Post, 13 Nov. 2005, http://www.washingtonpost.com/wp-dyn/content/article/2005/11/12/ AR2005111200958.html.
59. After having his DNA tested, the young man learned that a particular pattern on his Y-chromosome – passed from father to son – was also shared by two people with the same surname (distant relatives on his father's side). That surname, together with the place and date of birth of his father, was enough to track him down.
60. M. Gymrek, A. L. McGuire, D. Golan, E. Halperin and Y. Erlich, 'Identifying personal genomes by surname inference', Science, vol. 339, no. 6117, Jan. 2013, pp. 321–4, https://www.ncbi.nlm.nih.gov/pubmed/23329047.
61. Currently, genetic tests for Huntington's disease are not available from any commercial DNA testing kits.
62. Matthew Herper, '23andMe rides again: FDA clears genetic tests to predict disease risk', Forbes, 6 April 2017, https://www.forbes.com/sites/ matthewherper/2017/04/06/23andme-rides-again-fda-clears-genetic-tests-to-predict-disease-risk/#302aea624fdc.

Cars

1. DARPA, *Grand Challenge 2004: Final Report* (Arlington, VA: Defence Advanced Research Projects Agency, 30 July 2004), http://www.esd.whs.mil/Portals/54/Documents/FOID/Reading%20Room/DARPA/15-F-0059_GC_2004_FINAL_RPT_7-30-2004.pdf.
2. *The Worldwide Guide to Movie Locations*, 7 Sept. 2014, http://www.movie-locations.com/movies/k/Kill_Bill_Vol_2.html#.WkYiqrTQoQ8.
3. Mariella Moon, *What you need to know about DARPA, the Pentagon's mad science division*, Engadget, 7 July 2014, https://www.engadget.com/2014/07/07/darpa-explainer/.
4. DARPA, *Urban Challenge: Overview*, http://archive.darpa.mil/grandchallenge/overview.html.
5. Sebastian Thrun, 'Winning the DARPA Grand Challenge, 2 August 2006', *YouTube*, 8 Oct. 2007, https://www.youtube.com/watch?v=j8zj5lBpFTY.
6. DARPA, *Urban Challenge: Overview*.
7. 'DARPA Grand Challenge 2004 – road to …', *YouTube*, 22 Jan. 2014, https://www.youtube.com/watch?v=FaBJ5sPPmcI.
8. Alex Davies, 'An oral history of the DARPA Grand Challenge, the grueling robot race that launched the self-driving car', *Wired*, 8 March 2017, https://www.wired.com/story/darpa-grand-challenge-2004-oral-history/ .
9. 'Desert race too tough for robots', BBC News, 15 March, 2004, http://news.bbc.co.uk/1/hi/technology/3512270.stm.
10. Davies, 'An oral history of the DARPA Grand Challenge'.
11. Denise Chow, 'DARPA and drone cars: how the US military spawned self-driving car revolution', *LiveScience*, 21 March 2014, https://www.livescience.com/44272-darpa-self-driving-car-revolution.html.
12. Joseph Hooper, 'From Darpa Grand Challenge 2004 DARPA's debacle in the desert', *Popular Science*, 4 June 2004, https://www.popsci.com/scitech/article/2004-06/darpa-grand-challenge-2004darpas-debacle-desert.
13. Davies, 'An oral history of the DARPA Grand Challenge'.
14. DARPA, *Report to Congress: DARPA Prize Authority. Fiscal Year 2005 Report in Accordance with 10 U.S.C. 2374a*, March 2006, http://archive.darpa.mil/grandchallenge/docs/grand_challenge_2005_report_to_congress.pdf.
15. Alan Ohnsman, 'Bosch and Daimler to partner to get driverless taxis to market by early 2020s', *Forbes*, 4 April 2017, https://www.forbes.com/sites/alanohnsman/2017/04/04/bosch-and-daimler-partner-to-get-driverless-taxis-to-market-by-early-2020s/#306ec7e63c4b.
16. Ford, *Looking Further: Ford Will Have a Fully Autonomous Vehicle in Operation by 2021*, https://corporate.ford.com/innovation/autonomous-2021.html.
17. John Markoff, 'Should your driverless car hit a pedestrian to save your life?', *New York Times*, 23 June 2016, https://www.nytimes.com/2016/06/24/technology/should-your-driverless-car-hit-a-pedestrian-to-save-your-life.html.
18. Clive Thompson, Anna Wiener, Ferris Jabr, Rahawa Haile, Geoff Manaugh, Jamie Lauren Keiles, Jennifer Kahn and Malia Wollan, 'Full tilt: when 100 per cent of cars are autonomous', *New York Times*, 8 Nov. 2017, https://www.nytimes.com/interactive/2017/11/08/magazine/tech-design-autonomous-future-cars-100-percent-augmented-reality-policing.html#the-end-of-roadkill.

19. Peter Campbell, 'Trucks headed for a driverless future: unions warn that millions of drivers' jobs will be disrupted', *Financial Times*, 31 Jan. 2018, https://www.ft.com/content/7686ea3e-e0dd-11e7-a0d4-0944c5f49e46.
20. Markus Maurer, J. Christian Gerdes, Barbara Lenz and Hermann Winner, *Autonomous Driving: Technical, Legal and Social Aspects* (New York: Springer, May 2016), p 48.
21. Stephen Zavestoski and Julian Agyeman, *Incomplete Streets: Processes, Practices, and Possibilities* (London: Routledge, 2015), p. 29.
22. Maurer et al., *Autonomous Driving*, p. 53.
23. David Rooney, *Self-guided Cars* (London: Science Museum, 27 Aug. 2009), https://blog.sciencemuseum.org.uk/self-guided-cars/.
24. Blake Z. Rong, 'How Mercedes sees into the future', *Autoweek*, 22 Jan. 2014, http://autoweek.com/article/car-news/how-mercedes-sees-future.
25. Dean A. Pomerleau, *ALVINN: An Autonomous Land Vehicle In a Neural Network*, CMU-CS-89-107 (Pittsburgh: Carnegie Mellon University, Jan. 1989), http://repository.cmu.edu/cgi/viewcontent.cgi?article=2874&context=compsci.
26. Joshua Davis, 'Say hello to Stanley', *Wired*, 1 Jan. 2006, https://www.wired.com/2006/01/stanley/; and, for more detail, Dean A. Pomerleau, *Neural Network Perception for Mobile Robot Guidance* (New York: Springer, 2012), p. 52.
27. A. Filgueira, H. González-Jorge, S. Lagüela, L. Diaz-Vilariño and P. Arias, 'Quantifying the influence of rain in LiDAR performance', *Measurement*, vol. 95, Jan. 2017, pp. 143–8, DOI: https://doi.org/10.1016/j.measurement.2016.10.009; https://www.sciencedirect.com/science/article/pii/S0263224116305577.
28. Chris Williams, 'Stop lights, sunsets, junctions are tough work for Google's robo-cars', *The Register*, 24 Aug. 2016, https://www.theregister.co.uk/2016/08/24/google_self_driving_car_problems/.
29. Novatel, *IMU Errors and Their Effects*, https://www.novatel.com/assets/Documents/Bulletins/APN064.pdf.
30. The theorem itself is just an equation, linking the probability of a *hypothesis*, given some observed pieces of evidence, and the probability of that evidence, given the hypothesis. A more comprehensive introductory overview can be found at https://arbital.com/p/bayes_rule/?l=1zq.
31. Sharon Bertsch McGrayne, *The Theory That Would Not Die: How Bayes' Rule Cracked the Enigma Code, Hunted Down Russian Submarines, and Emerged Triumphant from Two Centuries of Controversy* (New Haven: Yale University Press, 2011).
32. M. Bayes and M. Price, *An Essay towards Solving a Problem in the Doctrine of Chances. By the Late Rev. Mr. Bayes, F.R.S. Communicated by Mr. Price, in a Letter to John Canton, A.M.F.R.S.* (1763), digital copy uploaded to archive.org 2 Aug. 2011, https://archive.org/details/philtrans09948070.
33. Michael Taylor, 'Self-driving Mercedes-Benzes will prioritize occupant safety over pedestrians', *Car and Driver*, 7 Oct. 2016, https://blog.caranddriver.com/self-driving-mercedes-will-prioritize-occupant-safety-over-pedestrians/.
34. Jason Kottke, *Mercedes' Solution to the Trolley Problem*, Kottke.org, 24 Oct. 2016, https://kottke.org/16/10/mercedes-solution-to-the-trolley-problem.
35. Jean-François Bonnefon, Azim Shariff and Iyad Rahwan (2016), 'The social dilemma of autonomous vehicles', *Science*, vol. 35, 24 June 2016, DOI 10.1126/science.aaf2654; https://arxiv.org/pdf/1510.03346.pdf.

36. All quotes from Paul Newman are from private conversation.
37. Naaman Zhou, 'Volvo admits its self-driving cars are confused by kangaroos', *Guardian*, 1 July 2017, https://www.theguardian.com/technology/2017/jul/01/volvo-admits-its-self-driving-cars-are-confused-by-kangaroos.
38. All quotes from Jack Stilgoe are from private conversation.
39. Jeff Sabatini, 'The one simple reason nobody is talking realistically about driverless cars', *Car and Driver*, Oct. 2017, https://www.caranddriver.com/features/the-one-reason-nobody-is-talking-realistically-about-driverless-cars-feature.
40. William Langewiesche, 'The human factor', *Vanity Fair*, 17 Sept. 2014, https://www.vanityfair.com/news/business/2014/10/air-france-flight-447-crash.
41. Bureau d'Enquêtes et d'Analyses pour la Sécuritié de l'Aviation Civile, *Final Report on the Accident on 1st June 2009 to the Airbus A330-203 registered F-GZCP operated by Air France Flight AF447 Rio de Janeiro – Paris*, Eng. edn (Paris, updated July 2012), https://www.bea.aero/docspa/2009/f-cp090601.en/pdf/f-cp090601.en.pdf.
42. Ibid.
43. Langewiesche, 'The human factor'.
44. Ibid.
45. Jeff Wise, 'What really happened aboard Air France 447', *Popular Mechanics*, 6 Dec. 2011, http://www.popularmechanics.com/flight/a3115/what-really-happened-aboard-air-france-447-6611877/.
46. Langewiesche, 'The human factor'.
47. Wise, 'What really happened aboard Air France 447'.
48. Lisanne Bainbridge, 'Ironies of automation', *Automatica*, vol. 19, no. 6, Nov. 1983, pp. 775–9, https://www.sciencedirect.com/science/article/pii/0005109883900468.
49. Ibid.
50. Alex Davies, 'Everyone wants a level 5 self-driving car – here's what that means', *Wired*, 26 July 2016.
51. Justin Hughes, 'Car autonomy levels explained', *The Drive*, 3 Nov. 2017, http://www.thedrive.com/sheetmetal/15724/what-are-these-levels-of-autonomy-anyway.
52. Bainbridge, 'Ironies of automation'.
53. Jack Stilgoe, 'Machine learning, social learning and the governance of self-driving cars', *Social Studies of Science*, vol. 48, no. 1, 2017, pp. 25–56.
54. Eric Tingwall, 'Where are autonomous cars right now? Four systems tested', *Car and Driver*, Oct. 2017, https://www.caranddriver.com/features/where-are-autonomous-cars-right-now-four-systems-tested-feature.
55. Tracey Lindeman, 'Using an orange to fool Tesla's autopilot is probably a really bad idea', *Motherboard*, 16 Jan. 2018, https://motherboard.vice.com/en_us/article/a3na9p/tesla-autosteer-orange-hack.
56. Daisuke Wakabayashi, 'Uber's self-driving cars were struggling before Arizona Crash', *New York Times*, 23 March 2018, https://www.nytimes.com/2018/03/23/technology/uber-self-driving-cars-arizona.html.
57. Sam Levin, 'Video released of Uber self-driving crash that killed woman in Arizona', *Guardian*, 22 March 2018, https://www.theguardian.com/technology/2018/mar/22/video-released-of-uber-self-driving-crash-that-killed-woman-in-arizona.
58. Audi, *The Audi vision of autonomous driving*, Audi Newsroom, 11 Sept. 2017, https://media.audiusa.com/en-us/releases/184.

59. P. Morgan, C. Alford and G. Parkhurst, *Handover Issues in Autonomous Driving: A Literature Review. Project Report* (Bristol: University of the West of England, June 2016), http://eprints.uwe.ac.uk/29167/1/Venturer_WP5.2Lit%20ReviewHandover.pdf.
60. Langewiesche, 'The human factor'.
61. Evan Ackerman, 'Toyota's Gill Pratt on self-driving cars and the reality of full autonomy', *IEEE Spectrum*, 23 Jan. 2017, https://spectrum.ieee.org/cars-that-think/transportation/self-driving/toyota-gill-pratt-on-the-reality-of-full-autonomy.
62. Julia Pyper, 'Self-driving cars could cut greenhouse gas pollution', *Scientific American*, 15 Sept. 2014, https://www.scientificamerican.com/article/self-driving-cars-could-cut-greenhouse-gas-pollution/.
63. Raphael E. Stern et al., 'Dissipation of stop-and-go waves via control of autonomous vehicles: field experiments', *arXiv: 1705.01693v1*, 4 May 2017, https://arxiv.org/abs/1705.01693.
64. SomeJoe7777, 'Tesla Model S forward collision warning saves the day', *YouTube*, 19 Oct. 2016, https://www.youtube.com/watch?v=SnRp56XjV_M.
65. Jordan Golson and Dieter Bohn, 'All new Tesla cars now have hardware for "full self-driving capabilities": but some safety features will be disabled initially', *The Verge*, 19 Oct. 2016, https://www.theverge.com/2016/10/19/13340938/tesla-autopilot-update-model-3-elon-musk-update.
66. Fred Lambert, 'Tesla introduces first phase of "Enhanced Autopilot": "measured and cautious for next several hundred million miles" – release notes', *Electrek*, 1 Jan 2017, https://electrek.co/2017/01/01/tesla-enhanced-autopilot-release-notes/.
67. DPC Cars, 'Toyota Guardian and Chauffeur autonomous vehicle platform', *YouTube*, 27 Sept. 2017, https://www.youtube.com/watch?v=IMdceKGJ9Oc.
68. Brian Milligan, 'The most significant development since the safety belt', BBC News, 15 April 2018, http://www.bbc.co.uk/news/business-43752226.

Crime

1. Bob Taylor, *Crimebuster: Inside the Minds of Britain's Most Evil Criminals* (London: Piatkus, 2002), ch. 9, 'A day out from jail'.
2. Ibid.
3. Nick Davies, 'Dangerous, in prison – but free to rape', *Guardian*, 5 Oct. 1999, https://www.theguardian.com/uk/1999/oct/05/nickdavies1.
4. João Medeiros, 'How geographic profiling helps find serial criminals', *Wired*, Nov. 2014, http://www.wired.co.uk/article/mapping-murder.
5. Nicole H. Rafter, ed., *The Origins of Criminology: A Reader* (Abingdon: Routledge, 2009), p. 271.
6. Luke Dormehl, *The Formula: How Algorithms Solve All Our Problems ... and Create More* (London: W. H. Allen, 2014), p. 117.
7. Dormehl, *The Formula*, p. 116.
8. D. Kim Rossmo, 'Geographic profiling', in Gerben Bruinsma and David Weisburd, eds, *Encyclopedia of Criminology and Criminal Justice* (New York: Springer, 2014), https://link.springer.com/referenceworkentry/10.1007%2F978-1-4614-5690-2_678.

9. Ibid.
10. João Medeiros, 'How geographic profiling helps find serial criminals'.
11. Ibid.
12. '"Sadistic" serial rapist sentenced to eight life terms', *Independent* (Ireland), 6 Oct. 1999, http://www.independent.ie/world-news/sadistic-serial-rapist-sentenced-to-eight-life-terms-26134260.html.
13. Ibid.
14. Steven C. Le Comber, D. Kim Rossmo, Ali N. Hassan, Douglas O. Fuller and John C. Beier, 'Geographic profiling as a novel spatial tool for targeting infectious disease control', *International Journal of Health Geographics*, vol. 10, no.1, 2011, p. 35, https://www.ncbi.nlm.nih.gov/pmc/articles/PMC3123167/.
15. Michelle V. Hauge, Mark D. Stevenson, D. Kim Rossmo and Steven C. Le Comber, 'Tagging Banksy: using geographic profiling to investigate a modern art mystery', *Journal of Spatial Science*, vol. 61, no. 1, 2016, pp. 185–90, http://www.tandfonline.com/doi/abs/10.1080/14498596.2016.1138246.
16. Raymond Dussault, 'Jack Maple: betting on intelligence', *Government Technology*, 31 March 1999, http://www.govtech.com/featured/Jack-Maple-Betting-on-Intelligence.html.
17. Ibid.
18. Ibid.
19. Nicole Gelinas, 'How Bratton's NYPD saved the subway system', *New York Post*, 6 Aug. 2016, http://nypost.com/2016/08/06/how-brattons-nypd-saved-the-subway-system/.
20. Dussault, 'Jack Maple: betting on intelligence'.
21. Andrew Guthrie Ferguson, 'Predictive policing and reasonable suspicion', *Emory Law Journal*, vol. 62, no. 2, 2012, p. 259, http://law.emory.edu/elj/content/volume-62/issue-2/articles/predicting-policing-and-reasonable-suspicion.html.
22. Lawrence W. Sherman, Patrick R. Gartin and Michael E. Buerger, 'Hot spots of predatory crime: routine activities and the criminology of place', *Criminology*, vol. 27, no. 1, 1989, pp. 27–56, http://onlinelibrary.wiley.com/doi/10.1111/j.1745-9125.1989.tb00862.x/abstract.
23. Toby Davies and Shane D. Johnson, 'Examining the relationship between road structure and burglary risk via quantitative network analysis', *Journal of Quantitative Criminology*, vol. 31, no. 3, 2015, pp. 481–507, http://discovery.ucl.ac.uk/1456293/5/Johnson_art%253A10.1007%252Fs10940-014-9235-4.pdf.
24. Michael J. Frith, Shane D. Johnson and Hannah M. Fry, 'Role of the street network in burglars' spatial decision-making', *Criminology*, vol. 55, no. 2, 2017, pp. 344–76, http://onlinelibrary.wiley.com/doi/10.1111/1745-9125.12133/full.
25. Spencer Chainey, *Predictive Mapping (Predictive Policing)*, JDI Brief (London: Jill Dando Institute of Security and Crime Science, University College London, 2012), http://discovery.ucl.ac.uk/1344080/3/JDIBriefs_PredictiveMappingSChaineyApril2012.pdf
26. Ibid.
27. A slight disclaimer. The PredPol algorithm itself isn't publicly available. The experiment we're referring to here was conducted by the same mathematicians who founded PredPol, using a technique that matches up to how the proprietary software

is described. All the clues suggest they're the same thing, but strictly speaking we can't be *absolutely* sure.

28. G. O. Mohler, M. B. Short, Sean Malinowski, Mark Johnson, G. E. Tita, Andrea L. Bertozzi and P. J. Brantingham, 'Randomized controlled field trials of predictive policing', *Journal of the American Statistical Association*, vol. 110, no. 512, 2015, pp. 1399–1411, http://www.tandfonline.com/doi/abs/10.1080/01621459.2015.107 7710.

29. Kent Police Corporate Services Analysis Department, *PredPol Operational Review*, 2014, http://www.statewatch.org/docbin/uk-2014-kent-police-predpol-op-review.pdf.

30. Mohler et al., 'Randomized controlled field trials of predictive policing'.

31. Kent Police Corporate Services Analysis Department, *PredPol Operational Review: Initial Findings*, 2013, https://www.whatdotheyknow.com/request/181341/response/454199/attach/3/13%2010%20888%20Appendix.pdf.

32. Kent Police Corporate Services Analysis Department, *PredPol Operational Review*.

33. This wasn't actually PredPol, but a much simpler algorithm that also used the ideas of the 'flag' and 'boost' effects. See Matthew Fielding and Vincent Jones, 'Disrupting the optimal forager: predictive risk mapping and domestic burglary reduction in Trafford, Greater Manchester', *International Journal of Police Science and Management*, vol. 14, no. 1, 2012, pp. 30–41.

34. Joe Newbold, '"Predictive policing", "preventative policing" or "intelligence led policing". What is the future?' Consultancy project submitted in assessment for Warwick MBA programme, Warwick Business School, 2015.

35. Data from 2016: COMPSTAT, *Citywide Profile 12/04/16–12/31/16*, http://assets.lapdonline.org/assets/pdf/123116cityprof.pdf.

36. Ronald V. Clarke and Mike Hough, *Crime and Police Effectiveness*, Home Office Research Study no. 79 (London: HMSO, 1984), https://archive.org/stream/op1276605-1001/op1276605-1001_djvu.txt, as told in Tom Gash, *Criminal: The Truth about Why People Do Bad Things* (London: Allen Lane, 2016).

37. Kent Police Corporate Services Analysis Department, *PredPol Operational Review*.

38. PredPol, 'Recent examples of crime reduction', 2017, http://www.predpol.com/results/.

39. Aaron Shapiro, 'Reform predictive policing', *Nature*, vol. 541, no. 7638, 25 Jan. 2017, http://www.nature.com/news/reform-predictive-policing-1.21338.

40. Chicago Data Portal, Strategic Subject List, https://data.cityofchicago.org/Public-Safety/Strategic-Subject-List/4aki-r3np.

41. Jessica Saunders, Priscilla Hunt and John Hollywood, 'Predictions put into practice: a quasi-experimental evaluation of Chicago's predictive policing pilot', *Journal of Experimental Criminology*, vol. 12, no. 3, 2016, pp. 347–71.

42. Copblock, 'Innocent man arrested for robbery and assault, spends two months in Denver jail', 28 April 2015, https://www.copblock.org/122644/man-arrested-for-robbery-assault-he-didnt-commit-spends-two-months-in-denver-jail/.

43. Ibid.

44. Ava Kofman, 'How a facial recognition mismatch can ruin your life', *The Intercept*, 13 Oct. 2016.

45. Ibid.

46. Copblock, 'Denver police, "Don't f*ck with the biggest gang in Denver" before beating man wrongfully arrested – TWICE!!', 30 Jan. 2016, https://www.copblock.org/152823/denver-police-fck-up-again/.

47. Actually, Talley's story is even more harrowing than the summary I've given here. After spending two months in jail for his initial arrest, Talley was released without charge. It was a year later – by which time he was living in a shelter – that he was arrested for a second time. This time the charges were not dropped, and the FBI testified against him. The case against him finally fell apart when the bank teller, who realized that Talley lacked the moles she'd seen on the robber's hands as they passed over the counter, testified in court: 'It's not the guy who robbed me.' He's now suing for $10 million. For a full account, see Kofman, 'How a facial recognition mismatch can ruin your life'.

48. Justin Huggler, 'Facial recognition software to catch terrorists being tested at Berlin station', Telegraph, 2 Aug. 2017, http://www.telegraph.co.uk/news/2017/08/02/facial-recognition-software-catch-terrorists-tested-berlin-station/.

49. David Kravets, 'Driver's license facial recognition tech leads to 4,000 New York arrests', Ars Technica, 22 Aug. 2017, https://arstechnica.com/tech-policy/2017/08/biometrics-leads-to-thousands-of-a-ny-arrests-for-fraud-identity-theft/.

50. Ruth Mosalski, 'The first arrest using facial recognition software has been made', Wales Online, 2 June 2017, http://www.walesonline.co.uk/news/local-news/first-arrest-using-facial-recognition-13126934.

51. Sebastian Anthony, 'UK police arrest man via automatic face recognition tech', Ars Technica, 6 June 2017, https://arstechnica.com/tech-policy/2017/06/police-automatic-face-recognition.

52. David White, Richard I. Kemp, Rob Jenkins, Michael Matheson and A. Mike Burton, 'Passport officers' errors in face matching', PLOSOne, 18 Aug. 2014, http://journals.plos.org/plosone/article?id=10.1371/journal.pone.0103510#s6.

53. Teghan Lucas and Maciej Henneberg, 'Are human faces unique? A metric approach to finding single individuals without duplicates in large samples', Forensic Science International, vol. 257, Dec. 2015, pp. 514e1–514.e6, http://www.sciencedirect.com/science/article/pii/S0379073815003758.

54. Zaria Gorvett, 'You are surprisingly likely to have a living doppelganger', BBC Future, 13 July 2016, http://www.bbc.com/future/story/20160712-you-are-surprisingly-likely-to-have-a-living-doppelganger.

55. 'Eyewitness misidentification', The Innocence Project, https://www.innocenceproject.org/causes/eyewitness-misidentification.

56. Douglas Starr, 'Forensics gone wrong: when DNA snares the innocent', Science, 7 March 2016, http://www.sciencemag.org/news/2016/03/forensics-gone-wrong-when-dna-snares-innocent.

57. This doesn't mean that false identifications are out of the question with DNA – they do happen; it just means you have a weapon on your side to make them as rare as possible.

58. Richard W. Vorder Bruegge, Individualization of People from Images (Quantico, Va., FBI Operational Technology Division, Forensic Audio, Video and Image Analysis Unit), 12 Dec. 2016, https://www.nist.gov/sites/default/files/documents/2016/12/12/vorderbruegge-face.pdf.

59. Lance Ulanoff, 'The iPhone X can't tell the difference between twins', *Mashable UK*, 31 Oct. 2017, http://mashable.com/2017/10/31/putting-iphone-x-face-id-to-twin-test/#A87kA26aAqqQ.

60. Kif Leswing, 'Apple says the iPhone X's facial recognition system isn't for kids', *Business Insider UK*, 27 Sept. 2017, http://uk.businessinsider.com/apple-says-the-iphone-xs-face-id-is-less-accurate-on-kids-under-13-2017-9.

61. Andy Greenberg, 'Watch a 10-year-old's face unlock his mom's iPhone X', *Wired*, 14 Nov. 2017, https://www.wired.com/story/10-year-old-face-id-unlocks-mothers-iphone-x/.

62. 'Bkav's new mask beats Face ID in "twin way": severity level raised, do not use Face ID in business transactions', Bkav Corporation, 27 Nov. 2017, http://www.bkav.com/dt/top-news/-/view_content/content/103968/bkav%EF%BF%BDs-new-mask-beats-face-id-in-twin-way-severity-level-raised-do-not-use-face-id-in-business-transactions.

63. Mahmood Sharif, Sruti Bhagavatula, Lujo Bauer and Michael Reiter, 'Accessorize to a crime: real and stealthy attacks on state-of-the-art face recognition', paper presented at ACM SIGSAC Conference, 2016, https://www.cs.cmu.edu/~sbhagava/papers/face-rec-ccs16.pdf.

64. Ira Kemelmacher-Shlizerman, Steven M. Seitz, Daniel Miller and Evan Brossard, *The MegaFace Benchmark: 1 Million Faces for Recognition at Scale*, Computer Vision Foundation, 2015, https://arxiv.org/abs/1512.00596

65. 'Half of all American adults are in a police face recognition database, new report finds', press release, Georgetown Law, 18 Oct. 2016, https://www.law.georgetown.edu/news/press-releases/half-of-all-american-adults-are-in-a-police-face-recognition-database-new-report-finds.cfm.

66. Josh Chin and Liza Lin, 'China's all-seeing surveillance state is reading its citizens' faces', *Wall Street Journal*, 6 June 2017, https://www.wsj.com/articles/the-all-seeing-surveillance-state-feared-in-the-west-is-a-reality-in-china-1498493020.

67. Daniel Miller, Evan Brossard, Steven M. Seitz and Ira Kemelmacher-Shlizerman, *The MegaFace Benchmark: 1 Million Faces for Recognition at Scale*, 2015, https://arxiv.org/pdf/1505.02108.pdf.

68. Ibid.

69. MegaFace and MF2: Million-Scale Face Recognition, 'Most recent public results', 12 March 2017, http://megaface.cs.washington.edu/; 'Leading facial recognition platform Tencent YouTu Lab smashes records in MegaFace facial recognition challenge', Cision PR Newswire, 14 April 2017, http://www.prnewswire.com/news-releases/leading-facial-recognition-platform-tencent-youtu-lab-smashes-records-in-megaface-facial-recognition-challenge-300439812.html.

70. Dan Robson, 'Facial recognition a system problem gamblers can't beat?', *TheStar.com*, 12 Jan. 2011, https://www.thestar.com/news/gta/2011/01/12/facial_recognition_a_system_problem_gamblers_cant_beat.html.

71. British Retail Consortium, *2016 Retail Crime Survey* (London: BRC, Feb. 2017), https://brc.org.uk/media/116348/10081-brc-retail-crime-survey-2016_all-graphics-latest.pdf.

72. D&D Daily, *The D&D Daily's 2016 Retail Violent Death Report*, 9 March 2017, http://www.d-ddaily.com/archivesdaily/DailySpecialReport03-09-17F.htm.

73. Joan Gurney, 'Walmart's use of facial recognition tech to spot shoplifters raises privacy concerns', iQ Metrix, 9 Nov. 2015, http://www.iqmetrix.com/blog/walmarts-use-of-facial-recognition-tech-to-spot-shoplifters-raises-privacy-concerns.

74. Andy Coghlan and James Randerson, 'How far should fingerprints be trusted?', *New Scientist*, 14 Sept. 2005, https://www.newscientist.com/article/dn8011-how-far-should-fingerprints-be-trusted/.

75. Phil Locke, 'Blood spatter – evidence?', *The Wrongful Convictions Blog*, 30 April 2012, https://wrongfulconvictionsblog.org/2012/04/30/blood-spatter-evidence/.

76. Michael Shermer, 'Can we trust crime forensics?', *Scientific American*, 1 Sept. 2015, https://www.scientificamerican.com/article/can-we-trust-crime-forensics/.

77. National Research Council of the National Academy of Sciences, *Strengthening Forensic Science in the United States: A Path Forward* (Washington DC: National Academies Press, 2009), p. 7, https://www.ncjrs.gov/pdffiles1/nij/grants/228091.pdf.

78. Colin Moynihan, 'Hammer attacker sentenced to 22 years in prison', *New York Times*, 19 July 2017, https://www.nytimes.com/2017/07/19/nyregion/hammer-attacker-sentenced-to-22-years-in-prison.html?mcubz=0.

79. Jeremy Tanner, 'David Baril charged in hammer attacks after police-involved shooting', *Pix11*, 14 May 2015, http://pix11.com/2015/05/14/david-baril-charged-in-hammer-attacks-after-police-involved-shooting/.

80. 'Long-time fugitive captured juggler was on the run for 14 years', FBI, 12 Aug. 2014, https://www.fbi.gov/news/stories/long-time-fugitive-neil-stammer-captured.

81. Pei-Sze Cheng, 'I-Team: use of facial recognition technology expands as some question whether rules are keeping up', NBC *4NewYork*, 23 June 2015, http://www.nbcnewyork.com/news/local/Facial-Recognition-NYPD-Technology-Video-Camera-Police-Arrest-Surveillance-309359581.html.

82. Nate Berg, 'Predicting crime, LAPD-style', *Guardian*, 25 June 2014, https://www.theguardian.com/cities/2014/jun/25/predicting-crime-lapd-los-angeles-police-data-analysis-algorithm-minority-report.

Art

1. Matthew J. Salganik, Peter Sheridan Dodds and Duncan J. Watts, 'Experimental study of inequality and unpredictability in an artificial cultural market', *Science*, vol. 311, 10 Feb. 2006, p. 854, DOI: 10.1126/science.1121066, https://www.princeton.edu/~mjs3/salganik_dodds_watts06_full.pdf.

2. http://www.princeton.edu/~mjs3/musiclab.shtml.

3. Kurt Kleiner, 'Your taste in music is shaped by the crowd', *New Scientist*, 9 Feb. 2006, https://www.newscientist.com/article/dn8702-your-taste-in-music-is-shaped-by-the-crowd/.

4. Bjorn Carey, 'The science of hit songs', *LiveScience*, 9 Feb. 2006, https://www.livescience.com/7016-science-hit-songs.html.

5. 'Vanilla, indeed', *True Music Facts Wednesday Blogspot*, 23 July 2014, http://truemusicfactswednesday.blogspot.co.uk/2014/07/tmfw-46-vanilla-indeed.html.

6. Matthew J. Salganik and Duncan J. Watts, 'Leading the herd astray: an experimental study of self-fulfilling prophecies in an artificial cultural market',

Social Psychology Quarterly, vol. 74, no. 4, Fall 2008, p. 338, DOI: https://doi.org/10.1177/019027250807100404.

7. S. Sinha and S. Raghavendra, 'Hollywood blockbusters and long-tailed distributions: an empirical study of the popularity of movies', *European Physical Journal B*, vol. 42, 2004, pp. 293–6, DOI: https://doi.org/10.1140/epjb/e2004-00382-7; http://econwpa.repec.org/eps/io/papers/0406/0406008.pdf.

8. '*John Carter*: analysis of a so-called flop: a look at the box office and critical reaction to Disney's early tentpole release *John Carter*', *WhatCulture*, http://whatculture.com/film/john-carter-analysis-of-a-so-called-flop.

9. J. Valenti, 'Motion pictures and their impact on society in the year 2000', speech given at the Midwest Research Institute, Kansas City, 25 April 1978, p. 7.

10. William Goldman, *Adventures in the Screen Trade* (New York: Warner, 1983).

11. Sameet Sreenivasan, 'Quantitative analysis of the evolution of novelty in cinema through crowdsourced keywords', *Scientific Reports* 3, article no. 2758, 2013, updated 29 Jan. 2014, DOI: https://doi.org/10.1038/srep02758, https://www.nature.com/articles/srep02758.

12. Márton Mestyán, Taha Yasseri and János Kertész, 'Early prediction of movie box office success based on Wikipedia activity big data', *PLoS ONE*, 21 Aug. 2013, DOI: https://doi.org/10.1371/journal.pone.0071226.

13. Ramesh Sharda and Dursun Delen, 'Predicting box-office success of motion pictures with neural networks', *Expert Systems with Applications*, vol. 30, no. 2, 2006, pp. 243–4, DOI: https://doi.org/10.1016/j.eswa.2005.07.018; https://www.sciencedirect.com/science/article/pii/S0957417405001399.

14. Banksy NY, 'Banksy sells work for $60 in Central Park, New York – video', *Guardian*, 14 Oct. 2013, https://www.theguardian.com/artanddesign/video/2013/oct/14/banksy-central-park-new-york-video.

15. Bonhams, 'Lot 12 Banksy: Kids on Guns', 2 July 2014, http://www.bonhams.com/auctions/21829/lot/12/.

16. Charlie Brooker, 'Supposing … subversive genius Banksy is actually rubbish', *Guardian*, 22 Sept. 2006, https://www.theguardian.com/commentisfree/2006/sep/22/arts.visualarts.

17. Gene Weingarten, 'Pearls before breakfast: can one of the nation's greatest musicians cut through the fog of a DC rush hour? Let's find out', *Washington Post*, 8 April 2007, https://www.washingtonpost.com/lifestyle/magazine/pearls-before-breakfast-can-one-of-the-nations-great-musicians-cut-through-the-fog-of-a-dc-rush-hour-lets-find-out/2014/09/23/8a6d46da-4331-11e4-b47c-f5889e061e5f_story.html?utm_term=.a8c9b9922208.

18. Quotations from Armand Leroi are from personal communication. The study he refers to is: Matthias Mauch, Robert M. MacCallum, Mark Levy and Armand M. Leroi, 'The evolution of popular music: USA 1960–2010', *Royal Society Open Science*, 6 May 2015, DOI: https://doi.org/10.1098/rsos.150081.

19. Quotations from David Cope are from personal communication.

20. This quote has been trimmed for brevity. See Douglas Hofstadter, *Gödel, Escher, Bach: An Eternal Golden Braid* (London: Penguin, 1979), p. 673.

21. George Johnson, 'Undiscovered Bach? No, a computer wrote it', *New York Times*, 11 Nov. 1997.

22. Benjamin Griffin and Harriet Elinor Smith, eds, *Autobiography of Mark Twain*, vol. 3 (Oakland, CA, and London, 2015), part 1, p. 103.
23. Leo Tolstoy, *What Is Art?* (London: Penguin, 1995; first publ. 1897).
24. Hofstadter, *Gödel, Escher, Bach*, p. 674.

Conclusion

1. For Rahinah Ibrahim's story, see https://www.propublica.org/article/fbi-checked-wrong-box-rahinah-ibrahim-terrorism-watch-list; https://alumni.stanford.edu/get/page/magazine/article/?article_id=66231.
2. GenPact, *Don't underestimate importance of process in coming world of AI*, 14 Feb. 2018, http://www.genpact.com/insight/blog/dont-underestimate-importance-of-process-in-coming-world-of-ai.

Index

Endnotes in the index are denoted by an 'n' after the page number e.g. 'ambiguous images 211n13'

Index

Index

Index

Index

choosing between individuals and the population 111
in fifteenth-century China 81
Hippocrates and 80
magic and 80
medical records 102–6
neural networks 85–6, 95, 96, 219–20n11
in nineteenth-century Europe 81
pathology 79, 82–3
patterns in data 79–81
predicting dementia 90–2
scientific base 80
see also Watson (IBM computer)
Meehl, Paul 21–2
MegaFace challenge 168–9
Mercedes 125–6
microprocessors x
Millgarth 145, 146
Mills, Tamara 101–2, 103
MIT Technology Review 101
modern inventions 2
Moses, Robert 1
movies see films
music 176–80
choosing 176–8
diversity of charts 186
emotion and 189
genetic algorithms 191–2
hip hop 186
piano experiment 188–90
algorithm 188, 189–91
popularity 177, 178
quality 179, 180
terrible, success of 178–9
Music Lab 176–7, 179, 180
Musk, Elon 138
MyHeritage 110

National Geographic Genographic project 110
National Highway Traffic Safety Administration 135
Navlab 117
Netflix 8, 188
random forests 59

neural networks 85–6, 95, 119, 202, 219–20n11
driverless cars 117–18
in facial recognition 166–7
predicting performances of films 183
New England Journal of Medicine 94
New York City subway crime 147–50
anti-social behaviour 149
fare evasion 149
hotspots 148, 149
New York Police Department (NYPD) 172
New York Times 116
Newman, Paul 127–8, 130
NHS (National Health Service)
computer virus in hospitals 105
data security record 105
fax machines 103
linking of healthcare records 102–3
paper records 103
prioritization of non-smokers for operations 106
nuclear war 18–19
Nun Study 90–2

obesity 106
OK Cupid 9
Ontario 169–70
openworm project 13
Operation Lynx 145–7
fingerprints 145
overruling algorithms
correctly 19–20
incorrectly 20–1
Oxbotica 127

Palantir Technologies 31
Paris Auto Show (2016) 124–5
parole 54–5
Burgess's forecasting power 55–6
violation of 55–6
passport officers 161, 164
PathAI 82
pathologists 82
vs algorithms 88

breast cancer research on corpses 92–3
correct diagnoses 83
differences of opinion 83–4
diagnosing cancerous tumours 90
sensitivity and 88
specificity and 88
pathology 79, 82
and biology 82–3
patterns in data 79–81, 103, 108
payday lenders 35
personality traits 39
advertising and 40–1
inferred by algorithm 40
research on 39–40
Petrov, Stanislav 18–19
piano experiment 188–90
pigeons 79–80
Pomerleau, Dean 118–19
popularity 177, 178, 179, 183–4
power 5–24
blind faith in algorithms 13–16
overruling algorithms 19–21
struggle between humans and algorithms 20–4
trusting algorithms 16–19
power of veto 19
Pratt, Gill 137
precision in justice 53
prediction
accuracy of 66, 67, 68
algorithms vs humans 22, 59–61, 62–5
Burgess 55–6
of crime
burglary 150–1
HunchLab algorithm 157–8
PredPol algorithm 152–7, 158
risk factor 152
Strategic Subject List algorithm 158
decision trees 56–8
dementia 90–2

Index

Index

HANNAH FRY

is an Associate Professor in the mathematics of cities at University College London. In her day job she uses mathematical models to study patterns in human behaviour, and has worked with governments, police forces, health analysts and supermarkets. Her TED talks have amassed millions of views and she has fronted television documentaries for the BBC and PBS; she also hosts the long-running science podcast *The Curious Cases of Rutherford & Fry* with the BBC.

·